1992	Hui N. L E. Rose David S Cathy L

EDITED BY

1994	S.L. Kane, MD	2015	L. Martin, MD J. Dionne, MD E. Wong, MD C. Cox, MD A. Smrke, MD K. Lewis, MD A. Mazzetti, MD G. Mazzetti, MD S. Patterson, MD N. Zamiri, MD
1997	D.J. Treleaven, MD N.E. Gibson, MD	2017	A. Kalani, MD A. Smrke, MD I. Al Nabhani, MD D. Chu, MD C. Coschi, MD A. Huynh, MD X. Li, MD M. Lovett, MD K. Matusiak, MD K. Onizuka, MD C. Park, MD M. Wang, MD
2001	Mark D. Soth, MD Azim Gangji, MD Dan Perri, MD C. (Russel) Mao, MD Tricia Woo, MD Daniel Hackam, MD		
2006	Y.D. You, MD Menaka Pai, MD Raffaela Profiti, MD		
2010	A. Bojcevski, MD F. Bacchus, MD		
2013	J. Cheung, MD O. Levine, MD S. Ning, MD J. Dionne, MD L. Martin, MD		

This book belongs to:

Disclaimer: The advice provided is of a general nature and needs to be tailored to the individual adult patient. Every attempt has been made to the make the information accurate. Furthermore, many aspects of medicine are matters of expert opinion and standard practices that may vary between institutions. Hence, no one associated with this guide can be held liable for any damages directly or indirectly caused by its contents – "nothing can substitute for sound medical training, experience and common sense".

Foreword to the 1992 (First) Edition

Hui N. Lee, E. Rose Jeans, Cathy Le Feuvre, David Strang

This guide is meant primarily for medical students and junior housestaff while on call in the Internal Medicine Services within the McMaster Teaching Hospitals. Thus, we named it, Survival Guide. Its only aim is to be an accessible and practical guide during those lonely stressful hours, and as such has obvious limitations. The authors emphasize that this is not a substitute for clinical acumen, knowledge, or the SMR! Also, there is a paucity of evidence-based protocols for many emergency situations, and therefore may be considerable variability among published and favoured regimens for these common problems: if your staff person prefers a different method, please don't blame us!

We would like especially to thank the following people for their invaluable help in proofreading and contributions to the manuscript:

Drs. Michael Achong, Chris Allen, Euan Carlisle, Deborah Cook, Stephan Sauve, Subash Jalali, William Nolan, Lori MacDonald Claude Korfas, Paul Tanser, Marion Sternbach, Peter Powers, Roman Jaeschke, Mo Ali, Rose Giammarco, David Morgan, Jim Nishikawa, David Russell, Anne Cranney, Noel Wright, Colin Barnes, Serge Puksa, PA McLellan, Coleman Rotstein, Susan Fawcett

However, the final draft and any errors are our responsibility and we apologize in advance. If the reader/user has any suggestions or comments, please send them to the Residency Office.

Finally, we should mention that all proceeds from the sales of this Guide go towards the McMaster Core Internal Medicine Residency Program's educational and social events.

It is our hope that future year residents will continue with updating the Survival Guide.

Foreword to the 2017 Edition

Aashish Kalani, Alannah Smrke, Ibrahim Al Nabhani, Derek Chu, Courtney Coschi, Amanda Huynh, Xena Li, Maggie Lovett, Kristine Matusiak, Kristyne Onizuka, Casey Park and Michael Wang are proud to update the McMaster Internal Medicine Survival Guide for their peers.

Thank you to the staff physicians that dedicated their time to edit our updates: Drs Waleed Al Hazzani, Zain Chagla, Serena Gundy, Ghazaleh Kazemi, Khurram Khan, Christian Kraeker, Mitch Levine, Mark Matsos, Matt Miller, Wes Oczkowzki, Jackie Quirt, Muntasir Saffie, Deb Seigel, Steven Sovran, Alan Tanaguchi, Vik Tandon, Eric Tseng, Tricia Woo and Doug Wright. Thank you Xena Li for her hard work formatting the entire book, and finally, as always, to Jan Taylor for her support throughout this process.

TABLE OF CONTENTS

ACLS ALGORITHMS

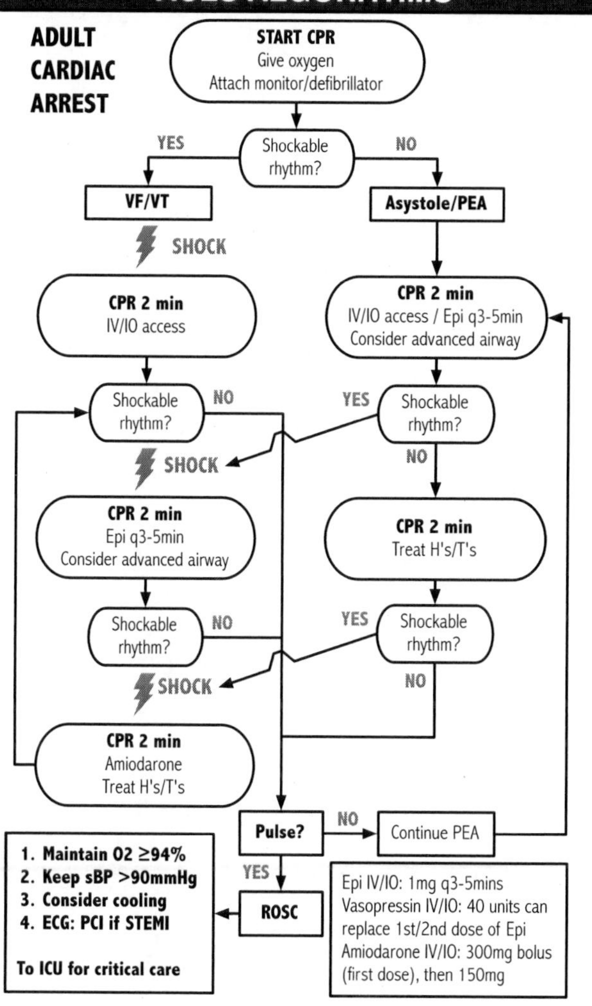

IMPORTANT

Reversible Causes:

H's	T's
• Hypovolemia • Hypoxia • Hydrogen ion (acidosis) • Hypo/hyperkalemia • Hypothermia	• Tension pneumothorax • Tamponade, cardiac • Toxins • Thrombosis, pulmonary • Thrombosis, coronary

- CPR – 2inches/5cm deep and fast (100-120bpm) and allow complete chest recoil
- Medications that can be given by ETT include Naloxone, Atropine, Vasopressin, Epinephrine, and Lidocaine (NAVEL)
  - Use 2-2.5 times the IV dose and dilute in 5-10 ml of NS.
- Intraosseus access can be obtained using the IO gun available on some crash carts

SYNCHRONIZED CARDIOVERSION

1. **Indications**: For unstable supraventricular tachycardias, AF/Flutter, Sustained VT with pulse
2. **Prepare equipment**: O2 monitor, Suction, IV, Intubation equipment.
3. **Sedation/analgesia if BP can tolerate** and call anaesthesia +/- respiratory therapist if time permits, if necessary, remove hair on chest to reduce impedence
4. Set energy levels
   - **Defibrillation**: 200J biphasic or 360J monophasic
   - **Cardioversion for A fib**: 120-200J (biphasic) or 200J (monophasic)with escalation PRN
   - **SVT NYD or atrial flutter**: 50-100J biphasic with escalation PRN.
5. If patient critically ill and machine does not synchronize then defibrillate.

TRANSCUTANEOUS PACING (TCP)

Important: TCP is only a temporary measure until reversible causes are remedied or a transvenous wire is placed

- Consider dopamine or epinephrine infusion while preparing for TCP (If low BP), Can also use Isoproterenol for symptomatic bardycardia as a bridge to TCP.
- Place pacing electrodes on chest and turn PACER on
- Set demand rate to 60/min then adjust to symptoms once capture achieved
- Set the current (mA) to ~10% (~5-10 mA) above the mA dose at which consistent capture is achieved (ie. every pacer spike is followed by QRS complex and every QRS complex is preceded by a pacer spike, also QRS complexes are coupled with a palpable pulse)
- Confirm electrical and mechanical capture (good pulse, blood pressure, symptom improvement)

ADULT BRADYCARDIA WITH PULSE

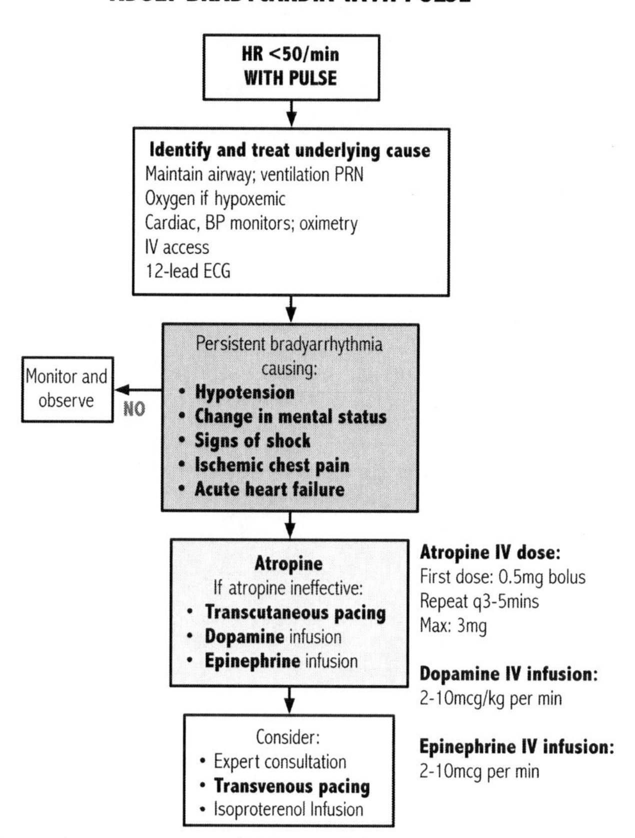

ADULT TACHYCARDIA WITH PULSE

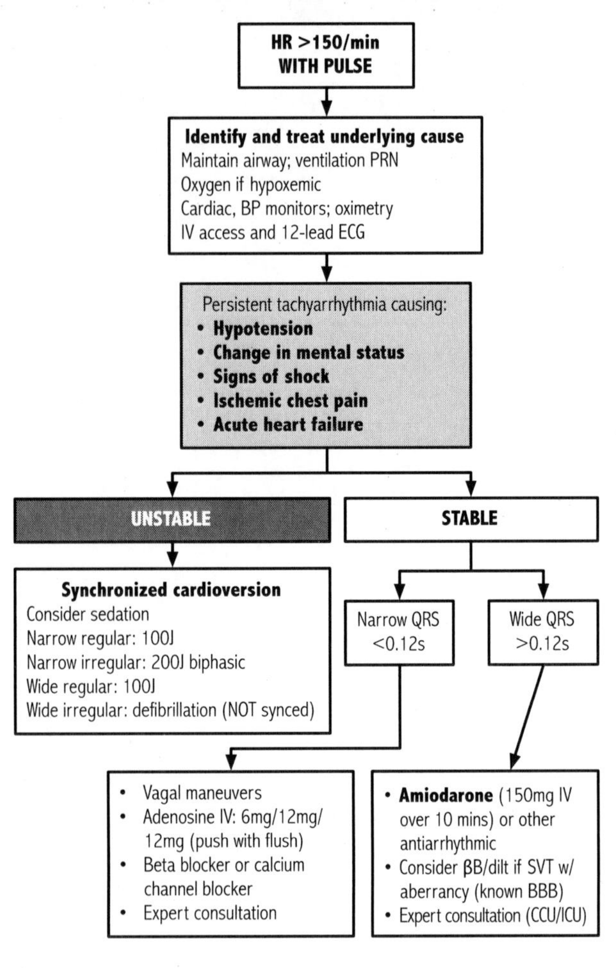

ACLS DRUGS

Adenosine: May terminate stable or unstable node-dependent SVT while awaiting cardioversion if necessary; give 6-12-12 mg IV push with saline flush q 5min.

Amiodarone: Non-arrest (stable VT, SVT with WPW, SVT if hypotensive, or wide-complex tachycardia NYD): Load 150 mg IV in 100 ml D5W in PVC/Glass, run over 10 min; may repeat q 10min prn; then 1 mg/min x 6 hrs (mix 900 mg in 500 ml D5W); then 0.5 mg/min x 18 hrs and beyond.
Arrest: 300 mg IV push (diluted in 20 ml D5W). May repeat 150 mg IV push x 1 in 3-5 min.
Maximum daily dose: 2.2 g IV.

Atropine: Unstable bradycardia: 0.5 mg IV push q3-5 min to max 3 mg.

Bicarbonate: Although little supportive evidence, bicarbonate may be considered in prolonged arrest and severe acidosis (metabolic, NOT for hypercarbic acidosis alone). Also consider in hyperkalemia, tricyclic OD, and for aspirin OD. Typical initial dose is 1 mEq/kg.

Calcium Chloride: In arrest or unstable patients with suspected or confirmed hyperkalemia; give 5-10 ml (10% solution, 0.5-1 g) slow IV push, may repeat PRN.

Diltiazem: First line Rx for afib or aflutter, and second line Rx for SVT (after vagal and adenosine)
Give 0.25 mg/kg (maximum first dose 15-20mg) IV push over 2 minutes, may repeat second dose of 0.35 mg/kg, then may infuse IV 5-15 mg per hour.

Dopamine: For cardiogenic shock dilute 400 mg in 250 ml D5W and infuse at 2-5 mcg/kg/min, titrating to target blood pressure (max 15mcg/kg/min).

Epinephrine: Arrest: 1 mg IV push q3-5 min or 2-2.5 mg diluted in 10 ml NS via ETT and consider continuous infusion by diluting 8mg in 250ml D5W or NS and infusing at 0.05-0.5 mcg/kg/min.
Profound bradycardia or hypotension: same infusion as above.

Magnesium sulfate: First line Rx for Torsades (hypomagnesemia, ischemia, bradycardia, long QT, medications)
Arrest: 1-2 g (2-4 ml of 50% solution) diluted in 10 ml D5W IV push
Non-arrest: 1-2 g hung in bag over 5-60 min.

Metoprolol: First line for SVT, afib or aflutter: Give 2.5-5mg IV push over 2 min (q5min prn to max of 15mg) or minibag infusion over 15 min

Morphine: For pain control and vasodilatation give 2-4 mg IV (over 1-5 min) q5-30min, to a maximum of 10mg/hr.

Norepinephrine: For septic shock, mix 8 mg in 250 ml D5W and infuse at 0.03-2 mcg/kg/min, titrating to BP

Procainamide: Suppression of stable monomorphic VT, afib, aflutter, WPW, or SVT in patients with preserved LV function: 20 mg/min IV infusion (max 17 mg/kg). Arrhythmia prevention: 1-4 mg/min IV infusion
Caution: Can cause hypotension or torsade.

Vasopressin: An alternative to epinephrine in cardiac arrest, but not on all arrest carts
May give 40 IU x 1 dose only (for pulseless VT/VF), but switch to Epinephrine if fails in 15 min.

Verapamil: First line Rx for afib or aflutter and second line Rx for SVT (after vagal and adenosine)
Give 2.5-5 mg IV over 2 min and repeat 5-10 mg IV q15-30 min to max 20 mg.

CCU/ICU Medication Dosing Guidelines

VASODILATORS		
Enalaprilat *(Vasotec)*	IV Bolus:	0.625 to 5 mg over 5 min then q6h (maximum 20 mg/day)
	Supplied:	1.25 mg/mL (2.5 mg/vial)
Hydralazine *(Apresoline)*	IV Bolus:	10-20 mg q4-6h
	Supplied:	20 mg vial
Nitroglycerin *(Tridil)*	IV Infusion:	10-200 mcg/min
	Supplied:	50 mg in total of 250 mL D5W
Nitroprusside *(Nipride)*	IV Infusion:	0.5-3 mcg/kg/min
	Supplied:	50 mg in total of 250 mL D5W

BETA BLOCKERS		
Esmolol *(Brevibloc)*	IV Bolus:	0.5 mg/kg over 1 min followed by infusion
	IV Infusion:	50 mcg/kg/min, ↑ by 50 mcg/kg/min q10min to maximum 200 mcg/kg/min Repeat bolus with each infusion increase
	Supplied:	Bolus: give bolus from infusion bag using bolus feature on the pump
	Infusion:	2500 mg in total of 250 mL D5W or NS
Labetalol *(Trandate)*	IV Bolus:	20 mg q10min (may need up to 100 mg)
	IV Infusion:	30-120 mg/h (0.5-2 mg/minute)
	Supplied:	Bolus: 5 mg/mL (100 mg/vial)
	Infusion:	100 mg in total of 100 mL **OR** 500 mg in total of 500 mL D5W or NS
Metoprolol *(Lopressor, Betaloc)*	IV Dose:	5 mg q3-5 min up to maximum of 15 mg, then 5-15 mg IV q6h
	Supplied:	1 mg/mL (5 mg/vial)

INOTROPES		
Dobutamine *(Dobutrex)*	IV Infusion:	2.5-20 mcg/kg/min
	Supplied:	250 mg in total of 250 mL D5W or NS
Dopamine *(Intropin)*	IV Infusion:	1-15 mcg/kg/min
	Supplied:	400 mg in total of 250 mL D5W
Milrinone *(Primacor)*	IV Bolus:	50 mcg/kg over 10 min followed by infusion
	IV Infusion:	0.25 -0.75 mcg/kg/min
	Supplied:	20 mg in total of 100 mL D5W or NS

VASOPRESSORS		
Epinephrine *(Adrenaline)*	IV Bolus:	1 mg
	IV Infusion:	0.05-0.5 mcg/kg/min
	Supplied:	Infusion: 8 mg in total of 250 mL D5W or NS Bolus: 1 mg/10 mL prefilled syringe
Norepinephrine *(Levophed)*	IV Infusion:	0.03-2 mcg/kg/min
	Supplied:	8 mg in total of 250 mL D5W or NS
Phenylephrine *(Neo-Synephrine)*	IV Bolus:	50-100 mcg (dilute 10 mg (1mL) in 100 mL IV bag to get 100 mcg/mL)
	IV Infusion:	0.05-2 mcg/kg/min
	Supplied:	10 mg/1 mL vial and 50 mg/5 mL vial Infusion: 50 mg in 250 mL D5W or NS
Vasopressin (Pitressin)	IV Infusion:	2 units/h
	Supplied:	20 units in total of 50 mL D5W or NS

Craig D Ainsworth. Adapted from J. Pickering.

CCU/ICU Medication Dosing Guidelines

DRUGS FOR DYSRHYTHMIAS		
Adenosine *(Adenocard)*	IV Bolus:	6 mg, repeat 12 mg in 2 min if necessary *(consider 3 mg initial dose if central line)*
	Supplied:	6 mg pre-filled syringe
Amiodarone *(Cordarone)*	IV Bolus:	150 mg or 300 mg in 100 mL D5W over 15 min, repeat in 30 min to total of 450 mg
	IV Infusion:	900 or 1200 mg/day (10-20 mg/kg/day)
	Supplied:	Central Line: 900 mg in 250 mL D5W Peripheral Line: 450 mg in 250 mL D5W
Atropine	IV Bolus:	0.5-1 mg, repeat if necessary q3-5min (maximum of 3 mg)
	Supplied:	1 mg pre-filled syringe
Digoxin (Lanoxin)	IV Loading:	0.5 mg, then 0.25 mg q4-6h x 2 doses then 0.125-0.25 mg q24h May require up to 1.5 mg for loading dose
	Supplied:	0.25 mg/mL (0.5 mg amp)
Diltiazem *(Cardizem)*	IV Bolus:	0.25 mg/kg over 2 min, if ineffective in 15-30 min give repeat bolus 0.35 mg/kg
	IV Infusion:	5-15 mg/h
	Supplied:	Bolus: 5 mg/mL (25 mg/vial) Infusion: 100 mg in 100 mL D5W or NS
Lidocaine	IV Bolus:	1-1.5 mg/kg over 2-3 min, may repeat Doses of 0.5-0.75 mg/kg q 5-10 min (maximum) 3 mg/kg.
	IV Infusion:	1-4 mg/min
	Supplied:	Bolus: 100 mg pre-filled syringe Infusion: 2 g in total of 500 mL D5W
Magnesium Sulfate	IV Infusion:	5 g at 1g/h (may give 2 g/h for first hour)
	Supplied:	vial: 5g in10 mL
	IV infusion:	5 g in total of 250 mL D5W/NS
Procainamide *(Pronestyl)*	IV bolus:	15-17 mg/kg (1000 mg) at 20 mg/min
	IV infusion:	2-6 mg/min
	Supplied:	Vial: 100 mg/mL (1g/10 mL vial) Infusion: 1 g in total of 250 mL D5W/NS

DRUGS FOR SEDATION IN VENTILATED PATIENT		
Fentanyl	IV Bolus:	25 –100 mcg (~ 0.35-1.5 mcg/kg)
	IV Infusion:	25 – 200 mcg/h (~ 0.7-10 mcg/kg/h)
	Supplied:	Bolus: 50 mcg/mL (2 mL amp) Infusion: 1000 mcg in 100 mL NS
Midazolam *(Versed)*	IV bolus:	1-5 mg (~ 0.02-0.08 mg/kg)
	IV infusion:	1-5 mg/h (~ 0.04-0.2 mg/kg/h)
	Supplied:	Bolus: 5mg/5mL, 10mg/2mL, 50mg/10mL Infusion: 50 mg in total of 50 mL D5W or NS
Morphine	IV Bolus:	1-5 mg
	IV Infusion:	1-5 mg/h
	Supplied:	Infusion: 100 mg in 100 mL D5W or NS
Propofol *(Diprivan)*	IV Bolus:	10-20 mg (caution hypotensive effects)
	IV Infusion:	0.3-3 mg/kg/h, increase by 0.3mg/kg/h q10 min until desired effect
	Supplied:	500 mg in 50 mL lipid base

Craig D Ainsworth. Adapted from J. Pickering.

ALLERGY AND IMMUNOLOGY

ANAPHYLAXIS

A rapid-onset and potentially life-threatening systemic hypersensitivity reaction

Common Causes: **Food** (peanuts, nuts, shellfish), **drugs** (Abx, NSAID+ASA, biologics, periop, chemo), insect stings radiocontrast, blood products, latex; less common: IV Fe^{2+}, hydatid cyst, SM/MCAS, spoiled fish, SPT, IT

Clinical features: Rapid onset/progression of Sx post exposure (mins-1 hr); Look for MedicAlert, check allergies

Dx (2006 NIAID/FAAN criteria): Depends on exposure + PmHx Allergy or not: Obvious Hx trumps criteria for Tx

Scenario	Criteria
Definite Hx allergy + allergen exposure	CVX only (ie. sudden onset of persistent low BP)
Likely Hx allergy + allergen exposure	2 or more of CVS/Resp/Derm/Mucosa Sx
NKA + allergen exposure	Derm/Mucosa Sx ***AND*** CVX or Resp Sx

Exception = delayed-onset anaphylaxis (hrs) to alpha-gal in cetuximab, ticks, or meat (esp beef)

In ER setting: sens 96.7%, spec 82.4%, PPV 68.6%, NPV 98.4%

CVS	Derm/Mucosa	HEENT/Ocular
• Low BP • Tachy-any type • Syncope • ACS/MI/CP • Arrest	• Urticaria • Angioedema • Erythema • Pruritus	• Lacrimation • Injection • Pruritus • Tongue swelling / drooling (angioedema)
Resp (upper and lower)	**GI**	**Neuro**
• SOB/Wheeze/cough • Resp distress, • Stridor/Hoarseness/ • Laryngeal edema • Rhinorrhea/Sneezing/ Congestion	• Abdo pain • N/V/D	• "Dizzy" • "Weak" • "feeling of doom" • Sz, AMS

Risk Factors that increase severity/mortality: BP or BB meds, CV or resp dz (esp asthma), no skin Sx, delay in Epi Tx, age

- Cofactors (combo allergen exposure + these can trigger a rxn): NSAID/ASA, exercise, alcohol, menstruation

NB: Death usually 2° to respiratory obstruction &/or CV collapse and within 60 mins

~5% (range 0.4% to 23%) biphasic rxn with recurrence of symptoms 1-72hrs later without further exposure

DDx: Depends on Sx/CC

- LOC – r/o hypo- or hyperglycemia
- CVS – SHOCK-E ddx (sepsis, hypovolemic/hemorrhagic, obstructive, cardiogenic, endocrine), etc. esp vasovagal
- Resp - Asthma exacerbation, COPDE, foreign body/aspiration, PE, panic
- ENT – Acquired/hereditary angioedema, oral allergy syndrome, rhinosinusitis, vocal cord dysfunction
- Rash – Broad ddx→precise morphology/distribution; acute vs chronic urticaria, vanco red man syndrome

Management: **If suspect anaphylaxis, give Epi STAT** – NO absolute contraindications (incl pregs, CV dz, ACS)

1. Remove trigger (ex. Stop drug/blood product infusion), supine or Trendelenburg position
2. **Epinephrine - give STAT when suspected**: give IM dose as 1st priority, then secure IV/IO access and airway, do not delay Epi for administration of antihistamines or steroids
   a. **0.3-0.5ml of 1:1000 Epi IM thigh q3-5min prn** (1mg/ml), should improve within mins, NB: 1:1000 for IM
   b. If severe hypotension: 0.5-1ml of 1:10,000 Epi (1mg/10ml) IV over 5 min (cardiac monitor req'd),
   consider epi infusion (ICU) if refractory to multiple boluses – bolus IV epi is associated with more complications over IM
3. IV O2 monitor +/- ETT (earlier the better esp w/ airway edema, bring difficult airway cart, consider calling ICU/Anesthesia for intubation), ABCs, Bolus IV crystalloid + prn, ?ICU
4. For resistant:
   a. Bronchospasm: salbutamol 2.5-5mg nebulizer, consider ipratropium, aminophylline
   b. Hypotension: Vasoressors – epinephrine, dopamine, norepinephrine, if refractory consider IV methylene blue
   c. Patient on a B-blocker/persistent low BP as other Sx resolve? → Glucagon 1 mg IV then q5min (max 5mg)
5. Reassess ABCs frequently (esp airway for edema), monitored setting, CCU/step down vs ICU
6. Optional adjunctive Tx - Antihistamines (for pruritus/urticaria)
   a. **Diphenhydramine (Benadryl) 25-50mg IV/po q4-6h** x 48-96hrs (max 400mg/24H) or Cetirizine (Reactine) 10-20mg PO q12-24H – this may sedate your patient; slow pharmacokinetics vs. rapidity of anaphylaxis
   b. **Ranitidine 50mg IV/150mg po** in addition may help pruritus
7. Optional adjunctive Tx - Corticosteroids: not effective for acute event but may prevent later recurrence
   a. **Methylprednisolone 125mg IV ,** then pred 50-100mg daily or methylpred 1-2 mg/kg//d q6h x48-96hrs
8. Observe closely for 6-8 hrs (biphasic reaction)
9. If no cardio/resp involvement (eg. only urticaria/rhinitis may trial Benadryl 25-50mg po/IV) with close monitoring + reassessment); otherwise give ***epi as it has no absolute contraindications***

Investigations: Only if idiopathic or insects→tryptase level r/o mastocytosis; IgE, etc falsely low after acute rxn

Follow-up:
- Identify+Avoid allergen/cofactor, Refer allergist, Rx epinephrine autoinjectors, MedicAlert, action plan
- Counsel patient; Refer Anaphylaxis Canada; Caution w/ BB/ACEi Rxs; If due to drugs, notify prescriber of drug;
- No good predictor of severity of next rxn (incl severity of last rxn)

Refs: WAO anaphylaxis guidelines 2015 update; US AAAAI/ACAAI JTF 2015 Practice Parameter

COMMON INPATIENT DRUG ALLERGIES

Common hypersensitivities to drugs: **Antibiotics, NSAIDs, ASA, Opioid, Radiocontrast**

General principles: Immediate (eg.anaphylaxis) vs delayed rxn (eg. SJS/TEN, DRESS, AIN, serum sickness)
- **Dx:** True allergy + confirmed Dx? Side effect of drug (eg. opioid N/V/pruritus/urticaria) vs. hypersensitivity
- **Tx:** Avoidance/discontinue, cross-reactivity, alternatives, graded challenge, desensitization, refer to allergist
- **Ix+F/U:** Confirmation of trigger – Hx; Skin prick testing (SPT) only useful for immediate hypersensitivities

Penicillin (PCN): IgE to part of β-lactam ring or R side chain (amox+amp); **also a common cause of delayed rxns**
- Self-report prevalence 10-20% 2° misconception of heritability, remote rxn, N/V/abdo pain, non-specific rash
  - True allergy prevalence is low: 10-20% of those who self-report (ie: ~1-4% overall are actually PCN allergic)
  - 80% patients outgrow/lose true penicillin allergy after 10 years
  - Hx is key to risk stratify; InpPt SPT useful and cost-effective in clarifying true immediate allergy or not
- B-lactam/lactamase inhibitors (NB: most are PCNs) – if Hx PCN anaphylaxis, then avoid or desensitize;
  - R/o allergy to lactamase inhibitor component
- Cephalosporins – ok in >96% PCN allergic; If ?PCN allergy, then SPT. If SPT impractical, graded challenge
  - If unlikely PCN allergy, then can receive full dose cephalosporin, esp 3rd or 4th generation
  - Exceptions: Amp allergy reacts w/ keflex (not ancef), cefaclor; Amox allergy cross-reacts with cefprozil
- Carbapenems – reported cross-reactivity <1% of individuals with +ve SPT to PCN; Graded challenge
- Aztreonam (monobactam) – no cross reactivity, can receive full dose

FQs: Causes either type of rxn; cross-rxn among all FQs → avoid FQ class or desensitize; SPT unreliable

Sulfonamides: Common: TMP-SMX, lasix, thiazides, acetazolamide, sulfonylureas, sulfasalazine, celecoxib, dapsone

- Can cause either immediate or delayed rxns; SMX associated with SJS/TEN and DRESS, not non-Abx sulfas
- Immediate allergy to SMX? Caution/graded challenge to sulfasalazine (mesalamine/5-ASA ok) – cross-rxn
  - Otherwise **does not** cross-react with non-Abx sulfonamides
- If Lasix allergy → can use ethacrynic acid instead, but greater risk ototoxicity

NSAIDs, including Aspirin: "6 types" of NSAID rxns due to either COX-1 inhibition or specific IgE to an NSAID

- Elicited by any NSAID incl ASA, ie. non-specific rxns/pseudoallergy; can be intermittent; **dose-dependent**
  - Type 1 – NSAID-induced asthma and rhinosinusitis; Sampter's triad = this + asthma + CRSwNP
  - Type 2 – Exacerbation of underlying chronic urticaria by NSAID
  - Type 3 – NSAID-induced urticaria/angioedema with no PmHx of chronic urticaria
  - Type 4 – Blended (mixed respiratory and/or cutaneous) reactions in otherwise asymptomatic individuals
- Specific to only one NSAID=true allergy: Type 5 and 6: Urticaria/angioedema vs. anaphylaxis, respectively
- If ACS/stroke – Use alternative agent (AHA 2013/2014); Can desensitize but can take ≥6 hrs
  - NB: Anaphylaxis (type 6 rxn) to ASA is rare; *Some* experts argue "no cases of anaphylaxis due to ASA";
- If Pain/arthritis/arthralgia – use non-NSAID analgesia/anti-inflammatory, COX-2 selective inhibitors OK
- Desensitization most useful for type I; eg. Pts with Sampter's triad who need outpt ASA CV prophylaxis

Opioids: Typically a non-specific rxn rather than a true allergy; can use cetirizine 10-20 mg po daily-bid

Radiocontrast:

- Premedicate per hospital policy (no uniform prophylaxis protocol);
  - Else, use low-osmolar contrast + Pred 50 mg @ 13, 7, 1 hr + Benadryl 50 mg po/IV 1 hr pre-procedure
- Shellfish/seafood allergy cross-reactivity to radiocontrast media is a misconception

Refs: ICON 2014 + US JTF 2010 guidelines, PCN allergy JAMA RCE (incl 2004 update), AAAI 2014:404, UpToDate

DRESS

Drug reaction with eosinophilia and systemic symptoms

Hypersensitivity reaction that starts 2-8 weeks after drug initiation, with frequent relapses despite cessation of drug trigger. HHV-6 viral infection/reactivation may play a role. Disease course generally lasts weeks-months.

Common Causes: Allopurinol, anticonvulsants (incl carbamazepine, phenytoin, lamotrigine), dapsone, septra, etc.

Clinical Features: Consider RegiSCAR criteria: Fever, LNs, facial edema, Eos, elevated LEs, AKI,

- Eosinophilia (>$0.4x10^9$/L), associated with leukocytosis and atypical lymphocytes
- Skin: morbilliform eruption progressing to a diffuse confluent lesions covering >50% BSA and follicular accentuation, may have facial edema, scaling, and purpura;
- Systemic symptoms: fever (>38°C), mild generalized lymphadenopathy, signs of end-organ dysfunction including elevated LEs (most common), liver failure, AKI (AIN). Less commonly lung (interstitial pneumonitis, hypoxia, cough, tachypnea/dyspnea), cardiac (esp Eos >1.5) any organ may be involved.
- Severe cases may be associated with: multi-organ failure, shock, DIC, or HLH

NB: Extent of rash (%BSA) correlates with severity - use Modified Lund-Bowder table

DDx: SJS/TEN, AGEP - <3 days post exposure, internal organs rarely involved; HES (not a drug trigger); lymphoma; acute cutaneous lupus erythematosus; erythroderma

Management:

- Withdrawal of triggering agent, avoidance in future
- Supportive care: skin moisture, emollients, fluid and nutritional support, management of pruritus with topical corticosteroids or oral antihistamines
- Systemic corticosteroids for organ dysfunction. 0.5-2mg/kg/day of prednisone until labs normalize, then taper over 8-12weeks. Relapses common while tapering.

STEVENS-JOHNSON SYNDROME/TOXIC EPIDERMAL NECROLYSIS

Life-threatening delayed drug rxn (4-28 days post exposure) causing desquamation with mucosal involvement

SJS (<10% BSA skin detachment), TEN (>30% BSA), overlap SJS/TEN (10-30% BSA)

Common Causes:
- Similar to DRESS; allopurinol, anticonvulsants; NSAIDs, sulfa abx; ? CMV or mycoplasma
- Epidemiology: women>men (0.6x M/F), can happen at any age. Increased rates in HIV infected individuals.

Clinical Features: Fever, painful desquamating skin with mucosal involvement, hypovolemic shock, hypernatremia
- General: Fever (>38°C), photophobia, odynophagia,. malaise, myalgia, arthralgia.
- Skin: Tender exanthematous eruptions that blister and slough; sparing palms, soles and scalp.
  - Nikolsky sign: superficial sloughing at an uninvolved site with gentle pressure
  - Asboe-Hansen (Bulla spread sign): lateral extension of bullae with gentle pressure
- Mucosal: Severe involvement (ocular, buccal, genital) with painful erosions, generally hemorrhagic and covered with greyish-white membrane. Conjunctival itching/burning/ corneal ulceration. Urethritis, genital erosions and urinary retention. May scar.
- Laboratory findings: lymphopenia/leukopenia. Eosinophilia or atypical lymphocytosis uncommon.
  - ± mild liver abnormalities of transaminases, (hepatitis in ~10%).
  - electrolyte abnormalities, hypoalbuminemia or increased BUN with severe cases

NB: Similar to severe burn patients with fluid/electrolyte abnormalities from environmental loss after skin sloughing (hypovolemic shock), secondary bacterial infection (*S. aureus/ P. aeruginosa*) causing septic shock.

DDx:
- Strep TSS, Staphylococcal scalded skin syndrome: spares mucous membranes, generally in very young
- Erythema multiforme: target lesions, limited to <10% detachment, HSV infection
- Erythroderma/exfoliative dermatitis: lack mucosal involvement and tenderness
- AGEP, vasculitis, pemphigus/pemphigoid, Phytotoxic eruptions: sun exposure and drug history

Management:

SCORTEN for dz severity on admission d1: if >2 burn unit/ICU, mortality 25%+
- Removal of triggering agent, avoidance in future
- Supportive care: +++fluid and electrolyte replacement, avoidance of bacterial infection, wound care, temperature management, nutritional support. Ocular care; ICU or burn unit
  - No role for prophylactic antibiotics
- No established therapy; systemic corticosteroids and IVIG are controversial

ATRIAL FIBRILLATION

Causes of Atrial Fibrillation:

Cardiogenic
- CAD/ACS
- CHF
- Valvular heart disease (including rheumatic heart disease) – MS, MR, TR
- Pericarditis/myocarditis
- HOCM
- Congenital heart disease

Pulmonary
- COPD
- OSA

- Age
- Genetics
- Hypertension
- Obesity
- Diabetes
- Metabolic syndrome
- Venous thromboembolic disease
- Post-surgical (cardiac or non-cardiac)
- Hyperthyroidism
- Infection/inflammation (i.e. lupus pericarditis)
- Alcohol/Caffeine

Acute Management:

General considerations
- If unstable (altered LOC, chest pain, low BP), proceed more urgently and cardiovert early (sedate and provide analgesia) (See ACLS algorithm for tachycardia)
- If there is a suggestion of Wolff-Parkinson-White (delta wave on resting EKG, young patient, HR>300 are suggestive of a bypass tract), avoid nodal suppression drugs (adenosine, beta-blocker, Ca-channel blocker, and digoxin) due to risk of ventricular arrhythmias and hypotension.
- Individuals considered particularly high risk of stroke include those with mechanical valves, prior stroke, and rheumatic heart disease

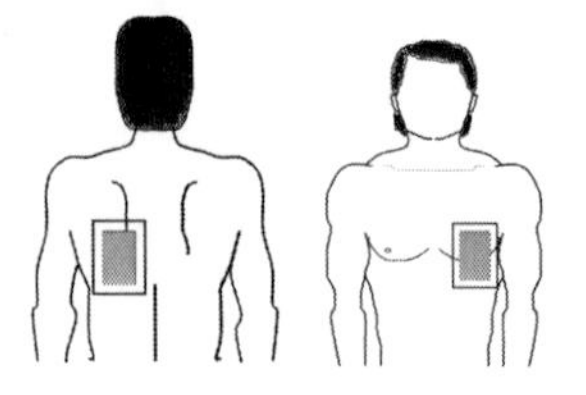

Paddle placement (Anterior posterior placement is better for maximal energy flow through atria)

Duration of Afib <48Hrs (and not high risk of stroke)

- **Unstable**: cardiovert immediately
  - (f patient is at high risk of stroke, then administer heparin, LMWH, Factor Xa inhibitor, or direct thrombin inhibitor as soon as possible before or immediately after cardioversion followed by long-term anticoagulation therapy
- **Stable**: pharmacologic or electrical cardioversion (if fails, proceed with rate control)
- **Initiate antithrombotic therapy according to CHADS2 score if Afib is persistent or recurrent**

Duration of Afib >48Hrs or Unknown Duration or High Risk of Stroke

- **Unstable**: cardiovert immediately
  - anticoagulate as soon as possible before or immediately after cardioversion and continue x4 weeks post cardioversion
- **Stable**: rate control
- If cardioversion is urgent, but there is time, a transesophageal echocardiogram is helpful to rule out left atrial or left atrial appendage thrombus
- If cardioverting, place on oral anticoagulation x 3 weeks before cardioversion and continue x 4 weeks post cardioversion
- Initiate longterm antithrombotic therapy according to CHADS2-VASC score

Acute Rate Control

Medication	Dose
Metoprolol	2.5-5mg IV over 2 min (max 15mg)
Diltiazem	0.25mg/kg (Max 20mg) over 10 min then repeat at 0.35mg/kg (Max 30mg)
Digoxin	0.5mg IV followed by 0.25mg IV Q6H x 2 for full loading dose

Acute Pharmacologic Cardioversion

Medication	Dose
Amiodarone	150mg IV over 10 min (may repeat x 1) then run 900mg over 24 hrs
Procainamide	15-17mg/kg IV over 60 min
Propafenone (in conjunction with AVnodal blockers)	450-600mg PO x 1

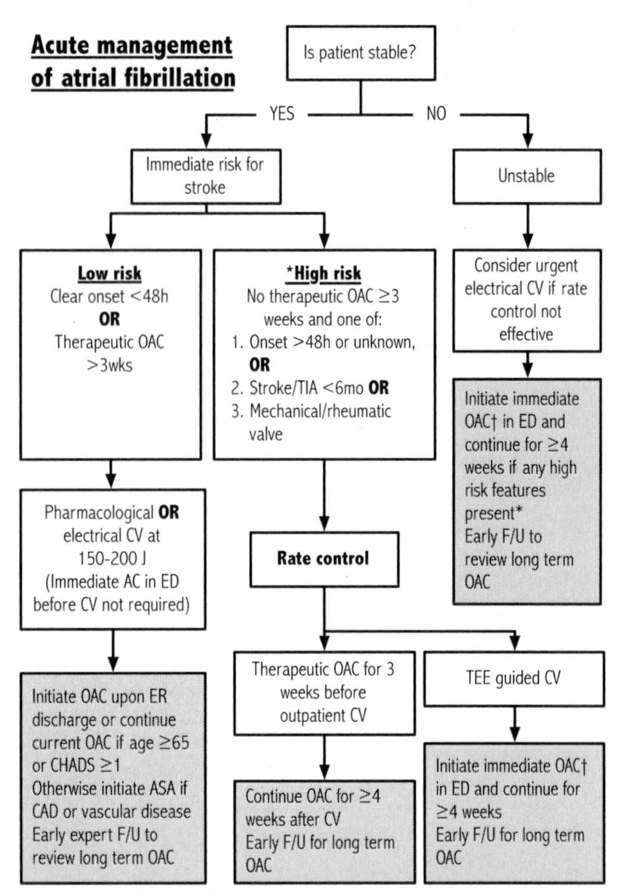

Decision algorithm for management of oral anticoagulation (OAC) therapy for patients who present to the emergency department (ED) with recent-onset atrial fibrillation (AF) requiring rate control or cardioversion (CV) in the ED. † Immediate OAC = a dose of OAC should be given just before cardioversion; either a NOAC or a dose of heparin or LMWH with bridging to warfarin if a NOAC is contraindicated. (2014 CCS Guidelines for management of AF).

Chronic Management:

General Considerations

- Investigate for underlying etiologies (TSH, extended lytes, CBC, troponins, echocardiogram)
- Address modifiable risk factors (caffeine, alcohol, smoking)
- Screen for other cardiac risk factors (hypertension, diabetes, dyslipidemia)

Rate Control

- Target resting heart rate <100bpm (CCS 2014 guidelines)
- Presence of CAD – beta-blocker (preferred) or calcium channel blocker
- Presence of CHF – beta-blocker +/- digoxin
- No CHF or CAD – beta-blocker or calcium channel blocker (add digoxin if uncontrolled)

Medication	Dose
Metoprolol	25-200mg PO BID
Bisoprolol	2.5-10mg PO daily
Atenolol	50-150mg PO daily
Diltiazem	120-480mg total daily dose
Digoxin	0.125-0.25mg PO daily (see loading dose above)

Rhythm Control

- Consider in patients with paroxysmal afib, new onset afib, age < 65, no history of hypertension, more symptomatic despite rate control
- Rhythm control does not offer any mortality benefit over rate control (AFFIRM trial)
- If EF<35%, amiodarone is the only recommended agent
- Dronedarone should not be used in permanent Afib (CCS 2012 guideline update)

Stroke Prevention: Stroke Risk Assessment: CHADS2 score

- CHADS2: CHF (non-valvular), Hypertension, Age (age >75), Diabetes, Stroke or TIA (2 points)
- CHA2DS2-VASc: CHF, HTN, Age≥75 (2 points), DM, Stroke or TIA (2 points), Vascular disease, Age 65-74 (1 point), Female Sex

CHADS2 Score	Annual Stroke Risk
0	0.6%
1	3%
2	4.2%
3	7.1%
4	11.1%
5	12.5%
6	13%

CHA2DS2-VASc Score	Annual Stroke Risk
0	0.2%
1	0.6%
2	2.2%
3	3.2%
4	4.8%
5	7.2%
6	9.7%
7	11.2%
8	10.8%
9	12.2%

Eur Heart J 2012; 33:1500. (Unadjusted for possible use of ASA)

CCS Algorithm for OAC Therapy in AF

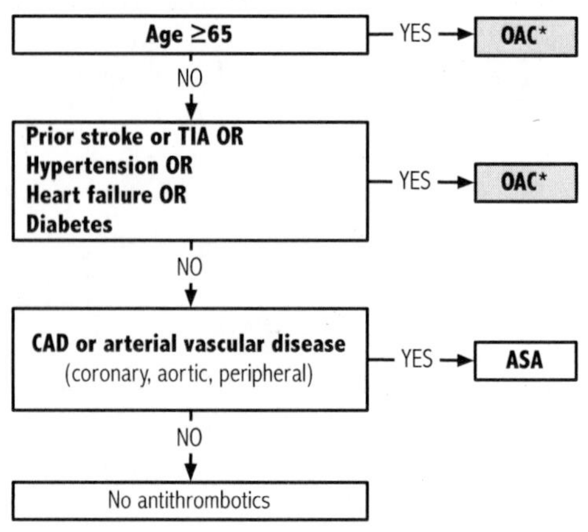

*Consider and modify all factors influencing risk of bleeding during OAC treatment (HTN, antiplatelets, NSAIDs, excess ETOH, labile INRs) an specifically bleeding risks for NOACs (low eGFR, age>75, low body weight).

Bleeding Risk Assessment: HAS-BLED Score

- HAS-BLED: Hypertension, Abnormal liver or renal function, Stroke, Bleeding, Labile INR, Elderly age (age >65), Drugs (use of drugs that may increase risk of bleeding, or alcohol use)
- Assesses the 1-year risk of major bleeding among patients with afib on oral anticoagulation

Antiplatelet Therapy:

ASA

- Appropriate for patients with afib, and history of CAD, or arterial vascular disease, but CHADS = 0, age <65

Combination with Clopidogrel

- Consider in patients whom vitamin K antagonists were deemed unsuitable (ACTIVE A trial)
- Although stroke risk was reduced with combination, the risk of bleeding increases significantly relative to ASA

Anticoagulation Therapy:
Warfarin
- Dose: Load 5-10mg daily x 3 days then reassess maintenance dose based on INR
- Target INR 2-3
- Reduces risk of stroke by 68% (Stroke 1997)

Dabigatran (RE-LY)
- Dose: 110mg BID or 150mg BID
- Stroke prevention: 110mg dose is non-inferior to warfarin and 150mg dose is superior to warfarin, less intracranial bleeding than warfarin for both
- Bleeding: less major bleeding with 110mg compared to warfarin and equivalent major bleeding with 150mg compared to warfarin
- 110 mg bid for patients ≥75 y/o due to better net clinical benefit (CCS 2014 Guidelines)

Rivaroxaban (ROCKET-AF)
- Dose: 20mg daily (if CrCl 30-49 reduce to 15mg daily)
- Stroke Prevention: non-inferior to warfarin
- Bleeding: less ICH, but more GI bleeds compared to warfarin

Apixaban (ARISTOTLE)
- Dose: 5mg BID (or 2.5mg BID if 2 or more of age ≥80, weight ≤60kg, Cr ≥133)
- Stroke prevention: apixaban is superior to warfarin in preventing ischemic or hemorrhagic stroke or systemic embolism
- Bleeding: less major bleeding with apixaban compared to warfarin with modest increase in survival
- Safe for CrCl≥30 ml/min

CONGESTIVE HEART FAILURE

Common Causes/Triggers

Cardiogenic
- MI –ECG, CK, trop, consider coronary angiography
- Valvular –regurgitation/stenosis-echo
- Pericardial: constriction, tamponade-echo
- Arrhythmias – bradycardia,;tachycardia- ECG
- Others: cardiomyopathy, septal defects, myocarditis, cardiac masses: Echo

Pulmonary
- PE,chronic pulonary disease (cor pulmonale): CXR, CTPE

Vascular/Renal
- Volume overload – iatrogenic (IVF), noncompliant to diet or meds (diuretics)
- Renal failure, renal artery stenosis
- Hypertensive emergency: see elsewhere

Metabolic, Infectious
- High output CHF (anemia, thyrotoxicosis, malnutrition, fistula etc.)

Infection

CXR Findings

- Vascular redistribution
- Interstitial edema ± Kerley B lines
- Increased cardiothoracic ratio (>0.5)
- Pleural effusions + fluid in the fissure
- Increased paratracheal stripe
- bronchial-alveolar cuffing

Management of Acute CHF:

1. ABC's, call SMR if patient unstable. Ensure HR/BP sufficient for diuresis
2. **L**asix, give intravenous initially and determine dose based on home lasix dose (i.e., double home dose), renal function, volume status etc. Monitor urine output carefully (can consider foley catheter; see Choosing Wisely). If inadequate diuresis, can consider increasing the dose (needs to be secreted into tubules, check renal fxn) or increasing frequency (pt may metabolize drug quickly) or trial of metolazone 2.5-10 mg po ½ hr prior to furosemide (thiazide-like diuretic; potentiates diuresis)
3. **M**orphine: venodilation (decreases preload) & decrease sensation of dyspnea; give 1-2 mg IV q10 min prn to max 10 mg/hr. Monitor LOC.
4. **NTG**: patch (0.4-0.8 mg/hr) or IV infusion (50 mg in 250 ml D5W start at 10 mcg/min) to decrease preload; make sure patient is not hypotensive and does not have AORTIC STENOSIS (preload dependent)
5. **O**xygen: apply high concentration. Call RT for **CPAP (PEEP decreases preload and afterload)**; intubation only if patient not maintaining oxygenation or continued elevated work of breathing
6. **P**osition: sit patient upright.
7. If patient is severely hypertensive, consider vasodilator (nitroprusside/glycerine) or enalaprilat.
8. Consider early use of ACE-I (short acting: captopril 12.5-25 mg po daily, decreases preload). Note: avoid starting any BB, and avoid nondihydropyradine CCB (ex. Diltiazem) in **acute** systolic heart failure.

Long-term Management of CHF:

Conduct a thorough history and exam to identify underlying etiologies

1. Perform investigations to find determine underlying etiology
   - BW: CBC, lytes, ext lytes, Cr/BUN, HbA1c, lipid profile. Perform TSH and other metabolic work up as clinically indicated (ex. Ferritin)
   - ECG to look for underlying ischemia/infarction, voltage, conduction abnormalities (etiology and treatment-CRT),
   - Echocardiogram for structural abnormalities and to estimate EF,
   - Coronary angiogram if concerns re: CAD, +/-MRI/cardiac biopsy/genetic screening
2. **Treat underlying etiology of CHF**
3. **Lifestyle measures**: diet, exercise, cholesterol/diabetes control, smoking cessation, reduction of alcohol consumption, decreased salt intake, fluid restriction

4. **ACEi/ARB**: mortality benefit, slows progression of LV dysfunction (HFrEF – heart failure with reduced EF)
5. **BB**: mortality benefit (HFrEF), slows progression; start at low dose and uptitrate as tolerated by BP. Avoid initiating in acute CHF due to negative inotropy
6. **Diuretics**: symptom control; usually furosemide, but can add metolazone to increase diuresis. Monitor for hypokalemia, renal function
7. **Aldosterone antagonists**: mortality benefit in CHF (HFrEF) (spironolactone/eplerenone), monitor for hyperkalemia, renal function
8. **Isosorbide dinitrate and hydralazine**: consider for those who are unable to tolerate ACEi/ARB because of intolerance, hyperkalemia, or renal dysfunction. (HFrEF)
9. **Antiplatelet/anticoagulation**: consider for those who have indications for these (i.e. CAD, afib, LV thrombus etc)
10. **Devices**: Cardiac Resynchronization Therapy (CRT) for primary prevention of arrhythmias and reduction in mortality if NSR, EF<35%, NYHA III/IV, QRS>120 or ICD if EF<35%
11. **Advanced therapies**: Mechanical support (ex. LVAD), inotropes (ex. milironone), transplantation for severe/refractory cases

ACUTE CORONARY SYNDROME (ACS)

Definitions:
- ***STE-ACS*** *(aka STEMI) (2014 ACCF/AHA Guidelines)* – new ST elevation at the J point in at least 2 contiguous leads of ≥2 mm in men or ≥1.5 mm in women in leads V2–V3 and/or of ≥1 mm in other contiguous chest leads or limb leads (new or presumably new LBBB has been considered a STEMI equivalent)
- ***NSTE-ACS*** *(formerly UA/NSTEMI) (2014 ACCF/AHA Guidelines)* – ST-segment depression or prominent T-wave inversion and/or elevated troponins in an appropriate clinical setting (chest discomfort or angina equivalent)
- ***Typical Angina*** *(not for acute presentations) (2012 ACCF/AHA Guidelines)* – 1) substernal chest discomfort with characteristic quality/duration; 2) provoked by exertion/emotional stress; and 3) relieved by rest/nitroglycerin
- ***Atypical Chest Pain*** – meets 2 of the above characteristics for typical angina
- ***Non-Cardiac Chest Pain*** – meets 1 or none of the above characteristics for typical angina

Initial Assessment:
1. **ABCs** – O2 to maintain sats ≥90%, IV access, monitor (frequent vitals and/or telemetry)
2. **ECG Review** – compare with old ECGs if available and, touch base with the Senior Medical Resident on-call if you are unsure of interpretation or if there are abnormalities, Always consider obtaining 15 lead ECG if high suspicion of ACS and normal ECG, obtain multiple ECGs to assess dynamic changes

Myocardium	ECG Leads	Vessels involved
Anterior/Septal	V1-V4, aVR*	LAD, PDA**
Lateral	V5-V6, I, aVL	LAD, Cx
Inferior	II, III, aVF	RCA, Cx
Posterior	V7, V8, V9	RCA, Cx
Right-sided	V4R, V5R, V6R	RCA

*ST elevation aVR (especially if STdepression in multiple leads): acute left main disease
** PDA: septal only (a branch of either Cx or RCA)

3. **History**
   - Timing – onset, duration, progression
   - Character – quality, quantity, location, radiation
   - Aggravating/Relieving – pleuritic, positional, post-prandial, nitro use, exertion, emotion
   - Associated Symptoms – N/V, diaphoresis, SOB, orthopnea, PND, palpitations, syncope, presyncope
   - Cardiac risk factors – HTN, DM, Smoking, Dyslipidemia, FHx of early CAD (males <55 or females <65)
   - ROS – URTI symptoms, cough, sputum, fever, constitutional symptoms, GI symptoms, trauma or muscle strain
   - Assess bleeding risk or history, need for upcoming surgeries and use or need for anticoagulation (DVTs, afib, mechanical valves, etc)

4. **Physical Exam**
   - Vital signs – check blood pressure in both arms, pulses in both hands
   - Cardio – check for murmurs, S3, S4, peripheral vascular exam
   - Resp – listen for wheezes and crackles
   - Volume status – check JVP and look for peripheral edema
   - Abdo – assess for tenderness to r/o intraabdominal causes of pain

5. **Investigations**
   - Serial ECGs with and without pain
   - Troponins Q3H x 2 (high sensitivity troponins)
   - 15 lead ECG if new R-wave and/or ST depressions in V1/V2
   - CXR for CHF
   - CBC, Cr, Lytes, Mg, PO4, Ca, Alb, LFTs, Check fasting glucose and lipids

6. **Differential diagnosis** – ACS, pulmonary embolus, pneumothorax, aortic dissection, pericarditis, esophageal rupture (Boerhaave's), perforated peptic ulcer, myocarditis, pneumonia, pleuritis, GERD, esophageal spasm, PUD, biliary, pancreatic, MSK, shingles, psychiatric

Management: Unstable Angina & NSTEMI
(Tip: remember MONA - morphine, oxygen, nitro, ASA)

1. **ABCs - O2, IV access, telemetry**

2. **Antiplatelet therapy**
   - ASA 160mg chewed x 1, then 81mg daily
   - Ticagrelor 180mg load and then 90mg bid (CCS 2012 guidelines on antiplatelets)

     OR
   - Plavix 300mg load x 1 (600mg if PCI planned), then 75mg daily (Do not use if on therapeutic anticoagulation)
   - Consider clopidogrel instead of ticagrelor if:
   - Fibrinolysis planned or given within 24 hours, Patient on chronic anticoagulation, Atrial Fibrillation, patient is high risk for bleeding
   - Dual anti-platelet should be continued for one year if patient medically managed
   - BMS inserted – dual antiplatelet for up to 1 year (at least 1 month)
   - DES inserted – dual antiplatelet for 1 year (at least 3 months if at high risk of bleeding) then continue beyond 1 year if risk of bleeding is acceptable

3. **Anticoagulation**
   - Start on admission; if PCI – continue until PCI procedure completed, if thrombolysis/no reperfusion strategy – continue until d/c from hospital

Class	Drug	Dose	Notes
Unfractionated Heparin (UFH)	Heparin IV	60units/kg bolus then 12	• Follow Low-dose Heparin

Indicated if going to cath lab w/in 24hrs, high risk of bleeding, CrCl<30 or need full dose AC (AF, mechanical valve, etc.)		units/kg/hr (max bolus 4000 units)	for ACS order set • safe in renal impairment • good option in moderate to high risk of bleeding
Factor Xa Inhibitor *(Preferred over LMWH with less bleeding and mortality (with thrombolysis)- OASIS 5 and 6 trial)*	Fondaparinux	2.5mg subcut daily	• contraindicated in CrCl <30ml/min
Low Molecular Weight Heparin (LMWH)	Enoxaparin	1mg/kg subcut Q12H (max 100mg per dose)	• contraindicated in CrCl <30ml/min

Note: If patient is on oral anticoagulation and INR is therapeutic and risk of bleeding is moderate to high OR if INR is supratherapeutic, do not give LMWH, UFH, Xa Inhibitor, or plavix.

4. Beta-Blockers

- Start metoprolol (starting dose 12.5 po BID) within 24 hours and titrate to achieve target HR 50-60
- IV beta-blocker no longer recommended. (Higher risk of mortality and cardiogenic shock)
- Avoid beta-blockers if worsening CHF, hypotension, bradycardia, risk for cardiogenic shock, PR > 0.24s, 2° or 3° AV block, severe asthma
- Has anti-anginal benefits for acute CP (decreases myocardial oxygen demand); could also consider CCBs

5. RAAS Inhibitors

- Initiate ACE-I within 24 hours of presentation (Ramipril 5mg daily, etc.)
- Start at low doses and titrate up as tolerated (monitor Cr and K)
- Particularly important in patients with anterior infarct, EF < 40%, CHF, or diabetic
- Do not give if SBP less than 100mmHg or less than 30mmHg below baseline
- If intolerant to ACE-I, may initiate ARB as an alternative

6. Statin

- Initiate high dose statin within 24 hours of presentation (ex. Atorvastatin 40-80mg po QHS)
- Check lipid levels within 24 hours of presentation
- Adjust statin dose accordingly

7. **Nitroglycerin**
- For symptom control in setting of ongoing angina
- Administer 0.4mg SL Q5min x 3 PRN then reassess for patch or drip
- If SL doses insufficient, may initiate patch (0.4-0.8 mg/h) or infusion (50 mg in 250 ml D5W, starting at 10 mcg/min, increase by 10 mcg/min Q5min until pain-free; max 150-200 mcg/min
- Do not administer if hypotensive or suspected/proven RV infarct and monitor for hypotension, tachycardia, and headache, don't administer if on PDE5 inhibitors (sildenafil, etc.) for the past 24-48 hrs

8. **Morphine**
- For symptom control in setting of ongoing angina
- Dose: 2-4 mg IV (over 1-5 min) Q5-30min, to a maximum of 10mg/hr

9. **Risk Stratification (invasive vs. conservative)**
- TIMI Score (low risk = score 0-2; medium risk = score 3-4; high risk = score 5-7)
- Early invasive angiography within 24 hrs if:

- High Risk NSTE-ACS ECG:
  - Transient ST elevation
  - ST elevation in AVR + diffuse ST depression
  - Posterior ST depression + no ST in V4R
  - New LBBB
- GRACE Score ≥140
- STEMI ≥12 hrs
- Refractory symptoms and/or dynamic ECG changes on medical therapy
- Hemodynamic instability or Killip 3,4 status
- ROSC with neurologic recovery
- Recurrent ventricular arrhythmias

- Conservative management can be considered if low TIMI (≤2) (or GRACE) score

TIMI risk score (add 1 pt for each)	TIMI CAD risk factors (≥ 3 out of 5)
• Age ≥ 65 years • ≥ 3 risk factors for CAD (see →) • Known CAD (stenosis ≥ 50%) • ASA use in past 7 days • Severe angina (≥ 2 episodes w/in 24 hrs) • ST changes ≥ 0.5mm* • Elevated serum cardiac markers (troponin/CK)*	1. Family history of CAD 2. Hypertension 3. Dyslipidemia 4. Diabetes 5. Current smoker

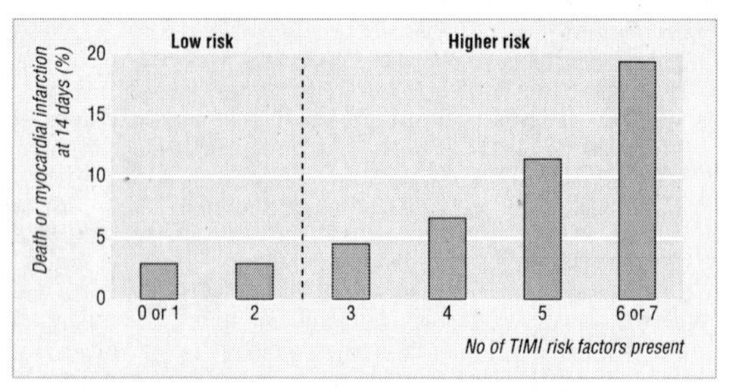

Rates of death from all causes and non-fatal myocardial infarction at 14 days, by TIMI risk score. Note sharp rate increase when score ≥3

BMJ. 2003 Jun 7;326(7401):1259-61.

*the presence of ST changes and Trop elevation generally warrants an invasive assessment (angiography) regardless of risk score

Management: STEMI:

1. Initial Management
   - Notify SMR immediately if suspected STEMI
   - SMR to contact interventional cardiologist on call if it is clearly a STEMI and patient a candidate for revascularization. The case should also be reviewed with Internal medicine staff on call.
   - Initiate management algorithm as above (follow same algorithm for NSTEMI)

2. **Percutaneous Coronary Intervention** (PCI)
   - Primary PCI if first medical contact (FMC) to device time is <90 min
   - Transfer for Primary PCI if FMC-device time <120min

3. **Thrombolysis (MUST DISCUSS with ER physician, internist, or cardiologist on call)**
   - Administer within 30 minutes if anticipated FMC-device time >120min
   - **Absolute contraindications** – any prior ICH, ischemic stroke or closed head trauma within 3 months, brain neoplasm or known structural cerebrovascular disease (eg. AVM), suspected aortic dissection, active bleeding or bleeding diathesis (except menses)
   - **Relative contraindications** – availability of PCI, INR > 2 on warfarin, history of severe chronic HTN or BP > 180/110 on presentation, major surgery within 3 weeks, traumatic or prolonged (>10 min) CPR, internal bleeding within 2-4 wks, uncorrected GI bleed, venous/arterial puncture to noncompressible site, ischemic stroke >3 month, dementia, any intracranial pathology, coma, pregnancy

Antithrombotic therapy for STEMI:

PCI	Thrombolysis
• IV Heparin bolus of 4000-5000 U as per direction from interventional cardiology	• IV Heparin Protocol as per pre-printed Thrombolysis order sets.
ASA 160mg load then 81 mg daily	
• Ticagrelor 180mg load then 90mg PO daily (preferred) **OR** • Plavix 600mg load then 75mg PO daily	• Plavix 300mg load (age <75) **OR** 75mg load (age ≥75) then 75mg PO daily

Administration of Thrombolytic Therapy:
- TNKase: single IV dose over 5s (<60kg=30mg; 60-69kg=35mg; 70-79kg=40mg; 80-89kg=45mg; >90kg=50mg), consider 50% dose reduction with age>75 yo (STREAM trial)
- Streptokinase: 1.5 million U over 30-60 min (contraindicated within 6 months of previous exposure because of potential inactivity)
- Note: Transient bradycardia is common in reperfusion therefore have atropine ready, but use only if bradycardia is severe and/or prolonged

Post-MI Complications:

Arrhythmias
- **AIVR** (accelerated idioventricular rhythm) is a hallmark of reperfusion and although ECG looks like slow VT, no treatment is necessary
- **PVCs and non-sustained VT**: check K, Mg (ideally keep K≥4.0 and Mg≥1.0) and pH and treat only if patient is symptomatic by using beta blockade, amiodarone, Mg or lidocaine (Class Ic anti arrhythmics are contraindicated- CAST trial)
- **Sustained VT or VF**: see ACLS algorithms

Hypotension/Shock
- **Potential causes**: LV or RV dysfnx, arrhythmias, acute MR from papillary muscle rupture (more common with inf. MI (PDA)), free wall rupture or ischemic VSD (septal rupture), tamponade, hematoma in post-CABG (tamponade physiology), iatrogenic (anaphylactic drug reaction), stent thrombosis
- **Investigations**: order urgent echocardiogram.
- **Management**: 250 ml NS if no evidence of pulmonary edema and follow with more volume if RV shock due to RV MI venodilation/hypotension due to inferior MI. If empirical volume not effective, patient may need Swan-Ganz catheter, IABP, inotropes (dopamine, norepinephrine, milrinone, dobutamine, epinephrine), mechanical support (LVAD)

Pericarditis – Three major types
- **Post-infarction pericarditis**: presence of a pericardial friction rub with or without C/P on day 1-2 post-MI (usually ECG changes overshadowed by MI findings). Do not stop anticoagulation (unless effusion develops). High-dose ASA (preferred) or colchicine. This is a poor prognostic factor.

- **Pericardial effusion**: check for pulsus paradoxus ± echo to rule out tamponade. and free wall rupture, treatment based on etiology/hemodynamics/symptoms..
- **Postcardiac injury syndrome (AKA Dressler's Syndrome):** a late complication with immune etiology usually develops weeks-months (rarely within a week) post MI. It is associated with pleuritic chest pain, a pericardial friction rub, fever, leukocytosis, high ESR, CRP, and sometimes, pleural effusion or pulmonary infiltrates. Treat with NSAIDs, eg. ibuprofen 400 tid-qid for 3-4 weeks with taper with colchicine ±steroids

Post-MI Management:

1. Lifestyle measures: diet, exercise, cholesterol/diabetes control, smoking cessation, reduction of alcohol consumption
2. Driving: restricted for 4 weeks post MI for private and 3 months for commercial
3. Repeat ECG/ECHO pre-discharge or within 1-3 months to assess new baseline and residual LV function.
4. **Antithrombotic therapy:**
    - ASA 81mg PO daily; usually indefinitely
    - Ticagrelor 90mg PO BID OR clopidogrel 75mg PO daily, ideally for at least 12 months (see above)
    - ± Warfarin/DOAC if history of Afib/VTE/LVthrombus – need to balance risk of bleeding
5. **ACEi/ARB:** prevents ventricular remodelling, definitely recommended for those the CHF EF <40%, but also for EF >40% (HOPE trial). ARBs if intolerant to ACEi
6. **BB:** start low and up-titrate as HR/BP tolerates (ie metoprolol 25mg PO BID)
7. **CCB:** not first line treatment. Consider as alternative to BB for angina (nondihydropyridine)
8. **Statin:** titrate to lipid profile (i.e. atorvastatin 40-80mg PO daily)
9. **Nitrates:** nitrospray 0.4mg SL q5mins x 3 prn for anginal pain. Use with caution in those with right-sided MI and who are preload dependent. Educate patients – if CP not resolving after at least 2 sprays – return to ED.
10. **Aldosterone antagonist:** EF <40% with DM or evidence of CHF
11. Referral to cardiologist if indicated for ongoing follow-up/monitoring and further risk stratification if worsening symptoms.

HYPERTENSIVE EMERGENCIES

Definitions:

1. **Emergency** = hypertension with end-organ damage:
    - **Encephalopathy**: cerebral edema, presented as severe H/A, vomiting, visual disturbances, transient paralyses, convulsions, stupor and coma
    - **Stroke**: ischemic or hemorrhagic (intracerebral or subdural hemorrhage)
    - **Retinopathy**: grade III (retinal exudate and hemorrhage) or grade IV (papilledema)
    - **Nephropathy**: acute renal failure, oliguria (<500 ml in 12 hours)
    - **Cardiovascular**: aortic dissection, acute pulmonary edema, UA/AMI
    - **Hematology**: microangiopathic hemolytoic anemia, schistocytes on peripheral smear

2. **Urgency** = asymptomatic, SBP > 200 or DBP > 130 ± optic disk edema, and peri-operative hypertension. It is not uncommon and is often due to non-compliance or refractory essential hypertension.
3. **Malignant HTN** is a historical definition of encephalopathy, nephropathy and papilledema.

Causes:

Most common: uncontrolled essential hypertension

- Renal vascular HTN (sudden rise in Cr when on ACEI or ARBs)
- Preeclampsia/ eclampsia/HELLP
- Pheochromcytoma
- Cushing's, 1° hyperaldosteronism
- Post head injury, stroke
- Aortic coarctation
- Drugs: cocaine (avoid betablockers), MAO-1, acute withdrawal of antihypertensives, non-compliance with BP medications, NSAIDs

Basic W/Us:

- CBC, lytes, BUN, Cr, urinalysis
- ECG/troponins (if CP), CXR
- When indicated, toxicology screen, pheo w/u (24 hr Urine metanephrines), eclampsia w/u, and CT head.

BP reduction strategies:

1. Treat the underlying cause (drug intoxication, metabolic/endocrine causes, ACS)
2. Urgency - use PO meds, BP must be reduced gradually over a few days; may manage on ward or as out-patient
3. Emergency - use IV meds, BP must be reduced within one hour; needs CCU/ICU
4. Don't aim for normal BP; reduce MAP [= (2 X DBP + SBP) / 3] by 10-20% in first hr and then by further 5-15% in next 23hrs. Goal is usually dBP<110 within 6 hrs.
5. Arterial line is essential for HTN emergencies.
6. Switch to oral maintenance therapy ASAP; a plan for long-term therapy should be introduced at the time of initial emergency treatment.
7. Ativan 1 mg S/L often useful adjunct.
8. Treat pain
9. If concerns re: pre-eclampsia/eclampsia – urgent OB/GYN consult.

Oral Agents for HTN Urgency:

	Dose	Kinetics	Cautions
Labetalol	200-400mg po q3-4hrs	Onset: 0.5-2hrs Peak: 3-4 hrs Duration: 6-8hrs	NOT in acute CHF, high degree heart block or asthma
Captopril	6.25-25mg po/sl q30-60min prn	Onset: 20 Min. Peak: 1hour Duration: 4-6hrs	Pregnancy, volume depletion, ARF, hyperkalemia, bilat. RAS
Clonidine	0.2mg po X1dose then 0.1mg po q1hr prn (max 0.8mg)	Onset: 30-60 mins Peak: 2-4 hrs Duration: 6-8hrs	Altered mental status, Watch for severe rebound HTN

IV Agents for HTN Emergency:

	Dose	Kinetics	Notes
Labetalol	20mg iv bolus then 10-40mg q10min or 0.5-2mg infusion (Max 300mg)	Onset: 5mins Peak: 10mins Duration 3-6hrs	Useful in most situations; especially Aortic Dissection
Hydralazine	5-20mg bolus, repeat q30min till target BP then q4hr	Onset: 10mins Peak: 30mins Duration: 2-6hrs	Can cause reflex tachycardia, not suitable in IHD and Aortic Dissection
NTG	5-200 mcg/kg/min infusion	Onset: 2-5 mins Duration: 5-10mins	Preferred in IHD and CHF; not as effective as other drugs in lowering BP
Nitroprusside	0.25-8mgcg/kg/min infusion	Onset: Seconds Duration: 3-5mins	Major SE: Cyanide poisoning Avoid Infusions >24-48hrs Increase toxicity with renal failure

Other IV agents: Nicardipine, Phentolamine, Fenoldapam

Specific Situations

- **SAH**: labetolol 1st choice, add nimodipine 60 mg po q4hr to decrease vasospasm; *avoid* nitroprusside and NTG.
- **ICH**: labetolol or nitroprusside.
- **Thromboembolic stroke**: avoid antihypertensives unless SBP > 220 or DBP > 120; labetolol 1st line, nicardipine 2nd line; *avoid* nitroprusside in general
- **Acute aortic dissection**: labetolol or nitroprusside + beta-blocker; *avoid* nitroprusside before adequate beta-blockage (HR < 60 beats/min). Target systolic BP: 100-120mmHg
- **Acute MI**: NTG, morphine and beta-blockers
- **Pulmonary edema**: NTG or nitroprusside, enalapril, morphine and loop diuretics
- **MAO-1 and tyramine**: phentolamine, labetolol, nitroprusside
- **Pheochromocytoma**: phentolamine and beta-blocker, phenoxybenzamine and beta-blocker, nitroprusside, labetolol; *avoid* beta-blockers before adequate alpha blockade.
- **Withdrawal of antihypertensives** (clonidine, propranolol): re-administer d/c'd drug
- **Scleroderma renal crisis**: ACEi

DIABETES: INPATIENT MANAGEMENT

General Principles

- Take a proactive approach to managing diabetes using basal, bolus, and supplemental insulin; avoid using sliding scales as exclusive management
- Non-critically ill: target preprandial CBG 5-8 mmol/L with random CBG <10mmol/L
- Critically ill: target preprandial CBG 8-10 mmol/L
- Perioperative: target CBG 5-10 mmol/L
- Avoid hypoglycemia (< 3.9 mmol/L) and severe hyperglycemia
- Patients with T1DM should be maintained on insulin at all times to prevent DKA
- If basal/bolus insulin: check CBG QID
- If NPO or continuous enteral feeding, check CBG Q4H
- If on IV insulin, check CBG Q1-2H
- Include orders to contact MD if CBG less than 4mmol/L or greater than 20mmol/L
- Generally, frequency of CBG checks are dependent on how well controlled sugars are

INSULIN INITIATION & MANAGEMENT

** ***REFER TO HHS ORDER SET: Adult Subcut Insulin Administration – Patient Eating /Adult Subcut Insulin Administration - Patient NPO*** **

Step 1: Calculate total daily dose (TDD) of insulin:

- General rule for DM1 and DM2: 0.5 units/kg/day

Step 2: Choose insulin regimen

1. Multiple Daily Injections (Optimal Choice)
   - 20% before each meal as Rapid or Short Acting
   - 40% at bedtime using NPH, Levemir or Lantus
   - Supplemental insulin to prevent excessive hyperglycemia and to guide adjustment of insulin doses prospectively
2. 2/3-1/3 Mixes:
   - Uses a combination of Intermediate and Rapid or Short Acting
   - Rarely started in hospital; if started, usually in conjunction with the DM nurse
3. Type 2 DM:
   - Patients with T2DM on oral medications with no contraindications to continue them in hospital (ex. AKI) can be started on bedtime insulin to improve management, and possibly meal-time insulin if needed
4. NPO patients:
   - Select appropriate fluid (if volume permits, D5W at 75-100mL/h or D10W at 30-50mL/h)
   - Use basal analogue (Lantus or Levemir) 0.3 Units/kg qHS or continue previous dose of long-acting insulin (consider reducing dose slightly if prolonged NPO)

Step 3: Reassess Daily (check on previous days CBG readings)

Multiple Daily Injections:

CBG	Too High (>11.1)	Too Low (<5.0)
Breakfast	↑ basal insulin 10%*	↓ basal 10%
Lunch	↑ breakfast insulin 10%	↓ breakfast insulin 10-20%
Dinner	↑ lunch insulin 10%	↓ lunch insulin 10-20%
Bedtime	↑ dinner insulin 10%	↓ dinner insulin 10-20%

*Consider checking a 3am CBG to assess for Somogyi Effect if persistent highs in AM despite increasing dose of long-acting insulin

Patient NPO – qHS basal analogue (Lantus or Levemir) regimen		
CBG	Too High (11mmol/l)	Too Low (5mmol/L)
Any time	Increase basal by 10% at HS	Decrease basal by 10-20% at HS

Step 4: Troubleshooting

1. LOW (blood glucose <4mmol/L)

- If patient conscious: give 15 g oral carbohydrate (2 x 114ml juice or 5 x 3g glucose tablets), repeat CBG Q15min, and give 15g carbohydrate every 15 minutes until CBG greater than 4mmol/L
- If patient unconscious or unable to eat: give 25ml D50W IV, repeat CBG Q15min, and give: 25 ml D50W every 15 minutes until CBG greater than 4 mmol/L

2. HIGH (blood glucose >20mmol/L)

- Give 0.15units/kg of short acting insulin (Humulin R, Novolin Toronto, Humalog, Novorapid) as one time dose or use sliding scale as a guide
- Re-check in 1-3 hours depending on peak time of insulin given
- Adjust standing insulin orders as per charts above, or using the sliding scale as a guide; DO NOT make adjustments based on one reading—look for trends

Step 5: Discharge Planning (refer to HHS Order Set)

- Preparation for discharge should begin 48 hours prior to expected discharge date
- All patients new to insulin therapy will require education related to injection of insulin, monitoring of capillary blood sugars, and nutrition
- Arrange dietetic or pharmacy consult as required
- Use Diabetes Canada Insulin prescription if desired (http://guidelines.diabetes.ca/CDACPG_resources/Insulin_Prescription_March_2014_update.pdf)

INSULIN PHARMACOKINETICS

INSULIN TYPE	ONSET	PEAK	DURATION
Bolus Insulins			
Rapid-acting (clear)			
• Insulin aspart (NovoRapid)	10-15 min	1-1.5 h	3-5 h
• Insulin glulisine (Apidra)	10-15 min	1-1.5 h	3-5 h
• Insulin lispro (Humalog)	5-15 min	1-2 h	3.5-4.75 h
Short-acting (clear)			
• Insulin regular (Humulin R)	30 min	2-3 h	6.5 h
• Insulin regular (Novolin Toronto)	30 min	2-3 h	6.5 h
Prandial Insulins			
Intermediate-acting (cloudy)			
• Insulin NPH (Humulin N)	1-3 h	5-8 h	Up to 18 h
• Insulin NPH (Novolin NPH)	1-3 h	5-8 h	Up to 18 h
Long-acting (clear)			
• Insulin detemir (Levemir)	90 min	No peak	16-24 h
• Insulin glargine (Lantus)	90 min	No peak	24 h

Canadian Diabetes Association Clinical Practice Guidelines 2013

General principles:
- Insulin is the preferred treatment for hyperglycemia (poor control) in hospitalized patients with diabetes
- Avoid oral agents in renal impairment and be mindful of other contraindications for specific agents
- If not reaching glycemic targets, optimize dose of current medications, add an agent from another class, or add/intensify an insulin regimen

Class	Relative A1C lowering	Hypo-glycemia	Weight	Other considerations
Acarbose	+	Rare	Neutral	Rarely used d/t GI side effects
Metformin	++	Rare	Neutral	Risk of lactic acidosis: do not use in renal failure, hepatic failure, IV contrast within past 24-48 hrs
Incretin agents: DPP-4 inhibitors	++	Rare	Neutral/↓	Caution with saxagliptin and CHF
GLP-1 agonists	+++	Rare	↓↓	GI side effects, high cost
Secretagogues: Meglitinide	++	Yes	↑	Less hypoglc but needs TID/QID
Sulfonylurea	++	Yes	↑	Gliclazide and glimepiride less hypoglycemia than glyburide
SGLT-2 inhibitors	++ / +++	Rare	↓	Avoid if eGFR <45-60mL/min, GU infections, normoglycemic DKA, CV benefits of empagliflozin
TZD (glitazones)	++	Rare	↑↑	Contraindicated w/ CHF, edema, fractures

Adapted from Canadian Diabetes Association Clinical Practice Guidelines 2013

DIABETIC KETOACIDOSIS (DKA)

***** REFER TO HHS ORDER SET: Diabetic Ketoacidosis Order Set for Adults *****

General Principles:

- Absolute or pronounced relative insulin deficiency; ketoacidosis predominates
- Hyperglycemia → urinary loss of water and lytes → volume depletion
- Ketoacidosis → K+ shift out of cells (overall total body potassium depletion)
- Resuscitate with aggressive IV fluids, IV insulin, and electrolyte (K+) replacement
- Focus should be placed on correcting the anion gap (AG) and maintaining normal electrolyte status (especially potassium)
- Identify and treat the **precipitant** (5I's – Infection, Ischemia, Intoxication, Initial presentation of type I DM, Insulin non-adherence)

1. ABCs

- Ensure patient has O2 as needed, IV access x 2, and is placed in a monitored setting
- VS Q15min x 4 then Q1hour x 4 then reassess

2. Initial Investigations

- Assess state of DKA and look for precipitant
- CBC, Cr, Lytes, HCO3, Ca, Mg, PO4, Alb, VBG or ABG, glucose, urinalysis, serum B-hydroxybuturate, osmolality
- 12-lead ECG, troponins, CXR, blood and urine cultures

3. IV Fluids

- DM1 pts can present with 6-8L of fluid deficit
- 0.9% NaCl 1L boluses, then high rate (250-500mL/h) and reassess hourly
- Once insulin is started, see K+ replacement for requirement in IVF

4. Potassium Replacement

- If anuric: hold KCl administration and reassess once urine output is established
- If K+ < 3.3mmol/L: add 40mEq KCl to IV fluids and hold IV Insulin until hypokalemia is corrected
- If K+ 3.3-5.0mmol/L: add 20mEq KCl to IV fluids
- If K+ >5.0mmol/L: hold KCl administration, start insulin drip, and reassess once K+ <5.0mmol/L

5. IV Insulin

- Dilute Regular Insulin 100 units in 100 ml of 0.9%NaCl (1 Unit/mL)
- Start at 0.1 Units/kg/hour and reassess rate based on anion gap
- Do not give a bolus of IV insulin
- Do not decrease IV insulin infusion rate if sugars normalize but anion gap and acidosis persists; instead manage decreasing glucose levels by adding D5W or D10W

6. Sodium Bicarbonate

- Consider NaHCO3 in patient with shock and/or arterial pH ≤ 7.0
- Do not use NaHCO3 if K < 3.3mmol/L (may exacerbate hypokalemia

7. Monitoring

- Bloodwork: CBGs hourly, electrolytes and VBG q2h x3 then R/A
- Monitor for correction of gap and acidosis, hypoglycemia, hypokalemia, hypophosphatemia
- Insert Foley catheter and monitor ins/outs hourly

8. Transitioning to Subcut Insulin

- Consider transitioning when AG is closed, acidosis corrected (pH ≥ 7.35), blood glucose ≤ 14mmol/L, IV Insulin rate < 2units/hour, patient is tolerating oral diet
- Overlap first subcut basal insulin dose with IV insulin infusion for 2-4 hours
- Restart patient on home regimen, or use above guidelines to initiate new insulin orders

HYPEROSMOLAR HYPERGLYCEMIC NON-KETOTIC COMA/SYNDROME (HONC/HHNS)

General Principles:

- Hyperosmolar and hyperglycemic; volume loss and hyperosmolarity predominate, minimal acid-base disturbance (vs DKA)
- Mortality is 25% in fulminant HHS
- Patients can have poorly controlled DM2 but an **underlying precipitant is far more likely**
    - o Search for precipitants like CVA, MI, CHF, infection, drugs (steroids, diuretics), post-op
- Signs and symptoms include severe dehydration (10-12L of fluid deficit) and altered mental status
- Bloodwork usually reveals severe hyperglycemia (glucose > 40-50), elevated serum osmolality (Osm>310mOsm/kg), no acidosis (pH >7.3), serum bicarbonate >15meq/L, normal anion gap, no ketones, hyperosmolar hyponatremia (see below)

Management:

1. **ABCs**
    - Ensure patient has O2 as needed, IV access x 2, and is in monitored setting
    - VS Q15min x 4 then Q1hour x 4 then reassess
2. **Initial Investigations**
    - CBC, Cr, Lytes, HCO3, Ca, Mg, PO4, Alb, VBG or ABG, glucose, urinalysis, B-hydroxybuturate, osmolality
    - 12-lead ECG, troponins, CXR, Blood and urine cultures
3. **IV Fluids**
    - Aggressive IV fluids until hemodynamic stability achieved and urine output >0.5ml/kg/hr
    - Start with 0.9% NaCl 1L over 30 min then reassess
    - Do not exceed plasma osmolality correction of more than 3 mOsm/kg/hr, or CBG correction greater than 5mmol/hr
4. **Insulin**
    - IV insulin still recommended to help reduce plasma glucose levels although less critical in HHS as there is no ketoacidosis to correct
    - If hyperglycemia persists despite adequate fluid resuscitation (ie. 4-5L of fluids, hemodynamic stability, and urine ouput 0.5ml/kg/hr), may start insulin drip at 1-2Units/hour
    - Do not give insulin if glucose < 14mmol/L
5. **Potassium Replacement**
    - If anuric: hold KCl administration and reassess once urine output is established
    - If K < 5.0mmol/L: add KCl to IV fluids (20-40mEq/L)
    - If K > 5.0mmol/L: hold KCl administration and reassess once K <5.0mmol/L
6. **Sodium Correction**
    - Remember to adjust Na levels based on glucose (for every ↑ in BG by 10 mmol/L correct Na by adding ~3 mmol/L)
    - Correct any hypoNa or hyperNa at a maxi rate of 8mEq/L over a 24 hour period
7. **Monitoring**
    - Bloodwork: Lytes, VBG, osmolality Q2h x3 then reassess
    - CBG hourly and insert foley catheter to monitor ins/outs

HYPOGLYCEMIA

General Principles:

- Etiology
  - o Drugs: insulin, insulin secretagogue, alcohol
  - o Critical illness: hepatic failure, sepsis
  - o Hormone deficiency: adrenal insufficiency/cortisol deficiency
  - o Endogenous hyperinsulinism: insulinoma, post-gastric-bypass hypoglycemia, autoimmune hypoglycemia
- Adrenergic signs and symptoms (mild): palpations, diaphoresis, tremor, hunger, anxiety, nausea, paraesthesias
- Neuroglycopenic signs and symptoms (moderate): agitation, confusion, bizarre behaviours, weakness, somnolence, visual and speech impairment, coma, seizure
- Whipple's triad (classically defined in insulinoma)

Workup:

During a hypoglycemic event send for the following blood tests:

- Glucose, insulin, C-peptite, beta-hydroxybutyrate, proinsulin
- Consider sending for sulfonylurea and meglitinide detection assays

Acute Management:

- **Mild-moderate**: give 15g oral carbohydrate*, repeat CBG Q15min, and give 15g carbohydrate every 15 minutes until CBG greater than 4mmol/L
- **Severe** (unconscious) with IV access: give 50ml D50W IV, repeat CBG Q15min, and give 50 ml D50W every 15 minutes until CBG greater than 4 mmol/L
- **No IV access and severe**: give glucagon 1mg IM/SC as a temporizing measure, then obtain IV access for glucose replacement as glucagon's effect will last only 30-45 minutes
- **Persistent hypoglycemia** (especially in sulfonylurea toxicity): may require a dextrose infusion (ex. D5W or D10W at 100ml/hr) with hourly CBG monitoring

**15g oral carbohydrate = 2 x 114ml orange juice OR 2 x 114 apple juice OR 1 x 114ml crancocktail or 1 x 114ml prune juice (or 25ml of D50W if unable to tolerate oral supplementation)*

HYPERCALCEMIA

Definition: corrected serum Ca > 2.6 mM.
- **Correct for hypoalbuminemia** or measure ionized calcium:
  - Formulas to correct for hypoalbuminemia do not apply for our assays at HHS/SJH

PTH mediated	PTH independent
1. **Primary Hyperthyroidism** • Parathyroid adenoma • Parathyroid hyperplasia • Parathyroid carcinoma 2. **Familial** • Familial Hypocalciuric Hypercalcemia • MEN1, MEN2A 3. **Tertiary Hyperparathyroidism** • CKD 4. **Lithium**	1. **Malignancy** • PTHrP (lung, renal, breast, bladder) • local osteolysis (breast, multiple myeloma) • cytokine release (leukemia, lymphoma) • metastases (direct invasion) 2. **Drugs** • Vitamin D, milk alkali syndrome, thiazides, • Tamoxifen, vitamin A, calcium 3. **Endocrine** • Addison's, hyperthyroidism, acromegaly (increased bone turnover), pheochromocytoma, adrenal insufficiency 4. **Granulomatous diseases** • sarcoidosis, TB, histo, Wegener's (due to increased 1 alpha – hydroxylase from macrophages, increased calcitriol) 5. **Others** • immobility, Zollinger-Ellison syndrome

Clinical Manifestations: "Bones, stones, groans, and psychiatric overtones"
- Often asymptomatic. Initial symptoms are fatigue, muscle weakness, inability to concentrate, nervousness, drowsiness.
- **CNS**: personality change, fatigue, depression, confusion/delirium, seizures, coma
- **GI**: anorexia, nausea, vomiting, constipation, abdo pain, pancreatitis
- **MSK**: bone pain, muscle weakness, hyporeflexia
- **CVS**: Short QT interval, increased digoxin effects
- **GU**: renal insufficiency, renal calculi, nephrocalcinosis, nephrogenic DI

Investigations:
- Blood – Ca, albumin, Mg, PO4, PTH, PTHrP, ALP, 25-hydroxy vitamin D, serum protein electrophoresis, CBC, lytes, BUN, Cr
- Urine – urine protein electrophoresis (multiple myeloma), 24 hour urine Ca, phosphorous and creatinine (for familial hypocalciuric hypercalcemia), 24 hour urine metanephrine (to rule out MEN2a)
- Imaging – neck and thyroid U/S, CXR and bone scan (for malignancy)
- ECG

Management:
- Hypercalcemia < 3.0 mM (corrected) rarely causes symptoms other than constipation and can usually be treated with fluids only

- **Treat underlying cause**
- Discontinue all medications that cause hypercalcemia (thiazide diuretics, Vitamin A, excess levothyroxine, lithium, excess active vitamin D)
- If symptomatic, or Ca > 3.0 mM, or ECG changes: **aggressive IV hydration** with NS boluses (1-2 litres) then maintain u/o 100cc/h with NS. Watch for volume overload.
- Weight-bearing mobilization of the patient if possible (immobilization aggravates hypercalcemia)
- Once patient is volume replete, may give **furosemide** 20-40 mg IV bid – tid.
- **Bisphosphonates**: Pamidronate 60-90 mg in 500 ml NS IV over 4 hours (adjust based on renal function). Zoledronate 4 mg in 50 ml NS IV over 15 minutes (more potent than pamidronate and can be given as a bolus). Takes 24-48 hours to have an effect. Calcium nadir in 5-7 days.
- **Calcitonin**: (NOT first line, most patients get bisphosphonate); Salmon calcitonin works rapidly however effect is short lived (tachyphylaxis within 48hrs); dose is 4u/kg/day (given in 2 doses) subcutaneous or IM or IV; ensure no allergy.
- **Corticosteroids**: Prednisone 10-40 mg po daily or equivalent, is sometimes helpful in malignancy and in hypervitaminosis D states (granulomatous diseases).
- **Cinacalcet** (calcimimetic) in hyperparathyroidism but rarely used.
- **Monitor electrolytes**, especially Ca, and Mg carefully.
- Hemodialysis in the rare patient with severe, symptomatic hyperCa (Ca >4.5-5) and neurologic Sx.

HYPOCALCEMIA

Definition: corrected serum Ca<2.1 mM
- As above, widely available corrections for hypoalbuminemia do not apply with our assays at HHS/SJH
- To confirm hypocalcemia, measure ionized calcium

Etiologies:

Low PTH	• Surgery, irradiation, autoimmune, infiltration, sepsis, hypomagnesemia • Hungry bone syndrome
PTH Resistance	• Pseudohypoparathyroidism
Low Vitamin D	• Malabsorption (IBD, diarrhea), poor nutrition, anticonvulsants (phenytoin: promotes inactivation of Vit D) • Reduced formation: Renal failure, liver failure
Complexation of Calcium	• Acute pancreatitis, rhabdomyolysis, tumor lysis syndrome, blood transfusions (citrate complexation), renal failure, chelators (EDTA, foscarnet, osteoblastic mets (esp breast, prostate)
Drugs	• Bisphosphonates, loop diuretics, calcitonin, phosphates, cinacalcet, ketoconazole, chemotherapy (5-FU, cisplatin)
Shift in Calcium Binding	• Alkalosis Metabolic or Respiratory (promotes Ca binding to albumin decreasing free Ca) • Hemodilution
Loss of Calcium	• Polyuric phase of renal recovery, diarrhea

Clinical Manifestations:

- **CNS**: confusion, irritability, depression, psychosis, increased ICP, seizures, diplopia
- **Neuromuscular**: hyperreflexia, cramps, perioral paresthesia, tetany, laryngospasm, Chvostek's (contraction of facial nerve with stimulus/tapping), Trousseau's (carpopedal spasms with inflation of BP cuff above systolic)
- **CVS**: prolonged QT, hypotension
- **GI**: steatorrhea
- **Renal**: osteodystrophy (low vit D, high PTH in renal failure) → osteomalacia, osteoporosis

Investigations:

- Blood – Ca, albumin, ionized calcium, Mg, PO4, PTH, 25 OH D3, ALP, lytes, BUN, Cr, 25 (OH)2D3, amylase, lipase, AST, ALT, total bilirubin, CK
- Urine Ca, creatinine
- ECG

Management:

- Treat the underlying cause
- Asymptomatic: do not need emergency treatment. May give oral Ca and Vit D supplementation
- If serum phosphate is elevated > 6 mM, consult Nephrology, concern for metastatic calcification with calcium replacement
- Two IV calcium preparations
  - 10% calcium chloride has 272mg elemental Ca/10mls
  - 10% calcium gluconate has 94mg elemental Ca/10mls
- If severe (Ca<1.5 mM) and symptomatic, give IV calcium gluconate 1 gram (10 cc of 10%) in 100 ml D5W and infuse over 30 mins.
  - If there is evidence of tetany or stridor, give 1 amp calcium gluconate IV over 2 mins. – and ensure cardiac monitoring, especially if on digoxin or have an arrhythmia
- If hypocalcemia persistent <1.9 mM or recurrent (eg. post renal transplant), may give calcium infusion. Mix 10 ampules of calcium gluconate 10% in 1L D5W or NS and infuse at 50-100 ccs/h, monitoring Ca q4-6h. Titrate by changing the rate of infusion.
- Can also start MgSO4 2g IV in 100 cc D5W over 2 hours (if low PTH is suspected)
- Post Para-thyroidectomy: mix 10 amps calcium gluconate in 1L D5W or NS and infuse at 50-100 ccs/h.
- In renal failure & hypoparathyroidism: Give 1,25-OH Vit D (calcitriol).
- Start oral replacement with calcium carbonate 500 -1000mg po tid (away from meals) +/- Vitamin D +/- thiazide diuretic; target calcium 1.9-2.1

CUSHINGS SYNDROME

Clinical Features:

- Hypokalemia, hyperglycemia, HTN, moon facies, supraclavicular fat pads, tanning, violet abdominal striae, osteoporosis, fragility fractures, abdominal obesity, thin skin, easy bruising

Investigations:

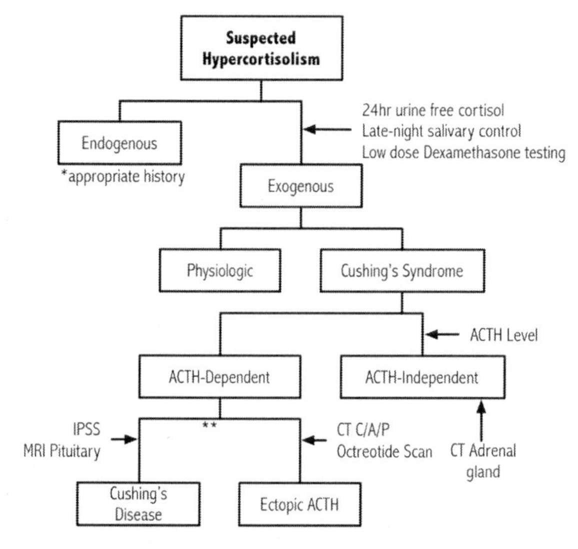

**Can consider HDDS, CRH or DDAVP stimulation tests to differentiate CD from ectopic-ACTH production, but their specificity is <90% and therefore not as useful

Acute Management:

- Treat hypercortisolism: ketoconazole, consider adrenalectomy, treat malignancy if the cause
- Supportive care:
  - Treat HTN with anti-hypertensives (avoid potassium-wasting diuretics)
  - Treat electrolyte abnormalities: spironolactone, potassium replacement
  - PCP prophylaxis

ADRENAL INSUFFICIENCY

Causes:

- Autoimmune, adrenal infiltration/trauma/hemorrhage, CAH, enzyme deficiency, ACTH-insensitivity, drug-induced (ex. Ketoconazole)

Clinical Features:

- Nausea and vomiting, abdominal pain, joint pain, salt craving, weakness, fatigue, poor appetite, weight loss, (pre) syncope, hyperpigmentation (if primary), ED, amenorrhea, fever

Objective Features:

- Hyperkalemia (if primary), hyponatremia, hypoglycaemia, orthostatic hypotension

Investigations

- AM cortisol, lytes can be suggestive
- ACTH stimulation test: baseline cortisol & ACTH levels followed by cosyntropin 250mcg IV x 1 then repeat cortisol level at 30 min and 60 min (cortisol < 450-500nM post-cosyntropin diagnostic of adrenal failure); can use 1mcg of cosyntropin as well

Treatment

- ABCs, IV fluids, monitored setting as required
- **Adrenal crisis** – hydrocortisone 100mg IV Q8H (preferable) or dexamethasone 4mg IV Q6H
  - Doses of hydrocortisone >100mcg/24h have sufficient mineralocorticoid activity
  - Dexamethasone is the only steroid formulation that will mostly not interfere with diagnostic tests for AI
- Chronic treatment – prednisone 5-7.5mg PO daily and fludrocortisone 0.05-0.1mg PO daily
- Patient education – medic alert bracelet and stress dose if unwell (double dose for 3 days)

COMPARISON OF COMMONLY USED CORTICOSTEROIDS

Glucocorticoid	Approximate Equivalent dose (mg)	Half-life (Biologic) (hours)
Short-Acting		
Cortisone	25	8-12
Hydrocortisone	20	8-12
Intermediate-Acting		
Methylprednisolone	4	18-36
Prednisolone	5	18-36
Prednisone	5	18-36
Triamcinolone	4	18-36
Long-Acting		
Betamethasone	0.6-0.75	36-64
Dexamethasone	0.75	36-64

HYPERTHYROIDISM

*** REFER TO HHS ORDER SET: Hyperthyroidism order set ***

Clinical Features:

- Heat intolerance, diaphoresis, tachycardia, hypertension, anxiety, fine tremor, hyperdefecation, weight loss, fatigue, light menses, palpitations, SOB, ophthalmopathy (Graves), pretibial myxedema (Graves), lid lag

Investigations:

- TSH, Free T4, Free T3, TRAb
- RAIU and Thyroid Scan
- Other blood tests as needed

Management Options: symptom relief, anti-thyroid medications, radioactive iodine, thyroidectomy

- Symptom relief
  - Volume replete first
  - Non-selective beta blockers: propranolol, metoprolol
  - If contraindicated (asthma, COPD, PVD, Raynaud's), then CCB: diltiazem, verapamil
  - Temperature control: acetaminophen and cooling blankets, avoid ASA (can exacerbate thyroid storm)
- Anti-thyroid medications (reduce thyroid hormone production)
  - Propylthiouracil 1000mg x 1 then 300mg Q6H then reassess in 48 hours (inhibits T3 → T4; safe in first trimester; first choice in thyroid storm)
  - Methimazole 30mg po Q6H then reassess in 48 hours (first choice in outpatients, more potent than PTU, longer lasting)
- May also need to administer lithium, Iodine, steroids, cholestyramine

Thyroid Storm:

- Severe and life-threatening hyperpyrexia, cardiovascular dysfunction, altered mentation with biochemical evidence of hyperthyroidism (Burch and Wartofsky score to diagnose and risk-stratify)
- Consult endocrinology
- Tx: beta blocker, PTU (preferred in thyroid storm), iodine solution, iodinated radiocontrast agent, glucocorticoids (relieve any adrenal insufficiency), bile acid sequestrants

HYPOTHYROIDISM

Clinical Features:

- cold intolerance, dry skin, bradycardia, delayed deep tendon reflexes, constipation, weight gain, anxiety, fatigue, myalgia, menstrual irregularities, tx for hyperthyroidism, goiter

Investigations:

- TSH, Free T4, Free T3
- Consider TPO Ab (if goiter) and thyroglobulin
- Other blood tests as needed

Management Options: symptom relief, normalize TSH, reduce goiter size, avoid overtreating

- Address constipation, warm the patient
- Levothyroxine 1.6mcg/kg/day; reduce starting dose in pts with CAD, elderly

Myxedema Coma (*high mortality):

- Severe hypothyroidism leading to decreased mental status, hypothermia and slowing of organ systems and metabolics; also consider hypoNa, hypoglc, hypothermia, hypoventilation
- Consult endocrinology
- Tx: thyroid hormone, supportive measures, manage precipitating problems; steroids first if possible adrenal insufficiency, rewarm with blankets

GASTROENTEROLOGY & HEPATOLOGY

ACUTE UPPER GI BLEEDING

Background:

- An upper GI bleed (UGIB) originates from any site proximal to the ligament of Treitz (attaches between the 3rd and 4th part of the duodenum)
- Main S&S are hematemesis, coffee ground emesis, melena or hematochezia, hemodynamic instability
- 90% of melena represents an UGI source, the remaining 10% comes from the remaining small intestine up to and including the right side of colon

Etiology:

- PUD (Physiologic Stress, NSAIDS, *H. pylori*, EtOH, smoking) (50%)
- Varices (15%)
- Esophagitis/gastritis/duodenitis (SSRI, oral bisphosphonates, GERD, EtOH, infections [HSV, candidiasis], radiation) (5-15%)
- Mallory Weiss tear (10%)
- Vascular malformations (eg. Dieulafoy, GAVE, aortoenteral fistula, Cameron lesions, angioectasia) (5%)
- Miscellaneous (portal hypertensive gastropathy, hemobilia, pseudohemobilia, orophryngeal bleed/epistaxsis)(5%)
- Upper GI malignancy (1-2%)

Diagnostic Approach:

- **HPI**: Acute or chronic, prior GIB, describe the bleed (hematemesis, coffee ground emesis, melena, BRBPR), constitutional symptoms, dysphagia, RF for esophageal cancers (smoking, hot liquids, obesity, reflux), RF for liver disease, EtOH consumption, hx vomiting/retching, abdo pain, AAA repair, bisphosphonates/SSRI/NSAIDS/anticoagulants, hx HTN, hx iron deficiency
- **NB**: bismuth, and iron may mimic melena stool
- **Exam**: Vitals (**include postural**), volume assessment, stigmata of chronic liver disease, CVS/Resp/Abdo, DRE
- **Lab**: CBC, Pt/PTT, BUN/Cr >36:1 suggests UGIB if no renal failure/diuretics; LFTs, lactate, Trops, EKG, Cross & type

Management:

1. Ensure ABCDE
    - **Assess hemodynamic stability → resuscitation 1st priority; endoscopy later**: Assess for shock, orthostatic hypotension, tachycardia, active bleeding, volume status
    - Step down/ICU if hemodynamically unstable, O2, Monitors
    - Prevent aspiration (prophylactic intubation may be required to protect airways in the setting of massive UGIB prior to endoscopy)
    - STAT two 16-18 gage peripheral IV's or central line if no peripheral access
    - NPO
    - Foley Catheter to urometer can help assess adequacy of resuscitation
    - AXR (3 views) if suspected perforation

2. **Emergency management of hypotension**
   - Decreased urine output (<0.5cc/kg/h), postural symptoms or orthostatic vitals (>20mmHg drop SBP OR >10mmHg drop DBP OR >30 beat increase in HR from lying to standing) suggests 20-40% of blood loss: bolus 1-2L RL ASAP
   - Supine tachycardia, hypotension, decreased LOC suggests >40% blood volume loss: bolus crystalloid ± blood and pressors until MAP>65 ASAP
   - Hold anti-hypertensive and diuretic medications
   - Call SMR who should call GI fellow and/or General Surgery.

3. **Blood transfusions (refer to *Hematology* chapter)**
   - PRBCs should be administered to patients without a cardiac history who have a Hb level less than 70 gms/L to maintain Hb 70-90.
   - If a patient has a cardiac history, aim should probably be to keep Hb >90.
   - Continue serial CBCs (including one post-transfusion) to watch for occult blood loss and response to transfusion (recall, 1U PRBC should raise your Hb level by 10 points, if it does not, then it implies that bleeding is continuous).
   - A massive transfusion is defined as needing >10U PRBCs in 24h or >5U PRBC in 3h
   - Give platelets if <20 and bleeding→ keep platelets >50

4. **Correct coagulopathies and discontinue antiplatelet, anticoagulants & NSAIDs**
   - INR ≥ 1.5 may serve as target threshold for correction of coagulopathy, but should not delay endoscopic therapy
   - Elevated INR → Vit K 2.5-10mg IV/SL (2.5-5mg if minor bleeding or 5-10mg if major bleeding at any INR elevation) and/or FFP 2-4U (or 15 ml/Kg) IV if rapid reversal needed
   - In life-threatening bleeding and elevated INR, call hematology to consider prothrombin concentrate complex (Octaplex),
   - If on IV heparin → D/C the infusion and give protamine infusion (1mg IV antagonizes 100U of heparin, max dose 50mg)

5. **Other Management**
   - Pantoprazole 80mg IV bolus then 8mg/hr x 72 hrs or 40mg IV/PO bid for variceal and non-variceal
   - If ? variceal bleed add octreotide 50mcg IV bolus then 50mcg/hr and ceftriaxone 1g IV q24h x7d
   - Use prognostic scale (Blatchford score) for early stratification of patients into low- and high-risk categories for rebleeding and mortality
   - Score = 0: low risk patient - do not require any intervention (blood transfusion, endoscopy, surgery)
   - Score > 6: associated with a > 50% risk of needing an intervention

Glasgow-Blatchford Score	
Admission risk marker	Score
Blood Urea	
≥6.5 <8·0	2
≥8.0 <10.0	3
≥10.0 <25.0	4
≥25	6
Hemoglobin (g/L) for men	
≥12.0 <13.0	1
≥10.0 <12.0	3
<10.0	6
Hemoglobin (g/L) for women	
≥10.0 <12.0	1
<10.0	6
Systolic blood pressure (mm Hg)	
100–109	1
90–99	2
<90	3
Other markers	
Pulse ≥100 (per min)	1
Presentation with melena	1
Presentation with syncope	2
Hepatic disease	2
Cardiac failure	2

Endoscopy:

- Does not need to be done in the middle of the night if patient stable. Should be performed within 24hr
- Consider erythromycin 250mg IV 30-60min before emergency EGD to improve visualization in patients with suspected blood in the stomach (i.e. bright red hematemesis)
- If bleeding recurs or patient is unstable call GI and/or Surgery early to "touch base." Little to no role for endoscopy if patient is hemodynamically unstable.

LOWER GI BLEEDS

Background:

- A lower GI bleed (LGIB) is one that originates from below the ligament of Treitz
- Main S&S are hematochezia, melena, occult bleeding, anemia, abdo pain
- 90% of melena represents an UGI source, the remaining 10% comes from below the ligament of Treitz up to and including the right side of colon

Etiology: (Causes are age-dependent)

- Structural: diverticulosis (painless & massive), haemorrhoids, anal fissure
- Vascular: angiodysplasia
- Infectious: *Campylobacter, E.coli, Salmonella, Shigella, Yersinia, C. difficile*
- Inflammatory: IBD (UC>>CD)
- Ischemic (think if patient is a known vasculopath with abdominal pain and weight loss)
- Neoplastic: polyp, carcinoma
- Others: radiation proctitis, brisk UGIB, post biopsy/polypectomy, Meckel's, ulcers, varices

Diagnostic Approach:

- **History**: acute or chronic, describe the bleeding (hematochezia (blood mixed in with stool), peri-rectal (eg. on tissue/outside of stool), melena, estimate blood loss, abdo pain, change in bowel habits/typical bowel habits, constitutional symptoms, infectious symptoms, syncope, recent antibiotic use, well-water use/camping, travel history, sick contacts, FHx IBD/colon cancer, anti-plt/anti-coag/NSAIDs/herbal
- **PMHx**: Previous GIB & C-scope, diverticulosis, hemorrhoids, IBD, CAD/PVD, radiation hx, malignancy, cirrhosis
- **Exam**: Vitals (**inc orthostatics**), volume exam, stigmata of chronic liver disease, CVS/Resp/Abdo, DRE + extraintestinal IBD screen (episcleritis/uvitis, oral ulcers, joint pain, pyoderma gangrenosum/erythema nodosum, perianal skin tags/fistulas, etc)
- **Lab**: CBC, Pt/PTT, BUN/Cr >36:1 suggests UGIB if no renal failure/diuretics; LFTs, Trops, EKG, Cross&type, lactate, iron studies, ± stool C&S, O&P, *C. diff* antigen

Management: See Steps 1-4 above in UGIB section

- Vitals, monitor $\pm O_2$ if unstable; **NPO** inititially
- 2x 16-18g IVs in AC
- CBC q2-12h → NB: initial Hb may be hemo concentrated and not correlate well with degree of blood loss; therefore check Hb post resuscitation and post transfusion.
- Conservative and supportive management initially
- Call GI and surgery early, these patients are generally older with more comorbidities
- May consider EGD as first test if massive UGIB suspected. Little role for colonoscopy if patient actively bleeding because poor visualization. Once bleeding settled colonoscopy is test of choice.
- If negative colo/EGD but ongoing bleeding, consider RBC scan (positive with 0.1cc / min) or angiography (positive with 0.5cc/ min).
- Selective Arterial Embolization (IR) after positive angiography used to coil vessels
- Small bowel studies including capsule endoscopy are indicated for obscure GI bleeds.
- Emergent exploratory surgery is last resort since all attempts should be made to try to localize bleeding site prior to surgery.

PANCREATITIS

There are 2 kinds of acute pancreatitis (AP):

1. **Edematous Pancreatitis→**Localized or diffuse enlargement of the pancreas secondary to edema that typically resolves within 1 week
2. **Necrotizing Pancreatitis→**A diffuse or focal area of non-viable pancreatic parenchyma that is >3cm in size or >30% (*NB:* May be pancreatic parenchymal and/or peripancreatic necrosis)

Definition of Acute Pancreatitis:

2 out of the 3 are needed to make the diagnosis of AP:

1. **Elevation of lipase and/or amylase three times the ULN.**
   - Lipase is more sensitive than amylase
2. **Abdominal pain consistent with AP**
   - Epigastric/LUQ abdominal pain that is constant and radiates to the back, better with sitting up
3. **Characteristic findings from abdominal imaging**
   - You do NOT need a contrast enhanced CT to confirm the diagnosis. Only order a CT if the diagnosis is unclear (i.e. abdo pain consistent with AP but lipase is not elevated)
   - Note: The full extent of necrosis will not be seen for 5-7days post-onset
   - An abdo U/S to rule out pancreatic gallstones should be ordered in everyone else

Assessment:

- **History**: Classify the abdo pain, gallstones/RUQ pain, jaundice/light stool/dark coloured urine, EtOH use, abdo trauma, PUD, constitutional symptoms, scorpion bites/travel, recent infections, autoimmune disorders, surgery/ERCP, dyslipidemia, hypercalcemia (moans, bones, groans, psychogenic overtones), medications/changes in medications.
- **Physical Exam**: vitals (inc postural if not too ill), volume assessment, abdo exam noting: loss of bowel sounds (paralytic ileus), Cullen/Grey Turner, Tetany, stigmata of liver disease, Resp/CV exam.
- **Labs**: CBC, lytes, Extended lytes, Creatinine/BUN, Lipase, BCx x2, IgG4, LFTs, EKG, abdo u/s vs. CT abdo for diagnosis + r/o gallstones, Lipid panel

Causes of Pancreatitis and Work-up:

Etiology	Work up
Idiopathic	All etiologies have been ruled out
Gallstones (causes an obstruction at the ampulla)	Laboratory: Specifically GGT, Alk phos, Bilirubin ±ALT, AST (<5x ULN) Imaging: Abdominal Ultrasound
Alcohol	Consider if: >50g/d and >5yr hx of EtOH consumption. Check for macrocytic anemia, AST:ALT >2:1

Trauma • Can be iatrogenic (ERCP)	History: Asking about any abdominal trauma
Medications Eg.: steroids, azathioprine, 5-ASA, 6-MP, Mesalamine, Valproic acid, ACEi, statins, antibiotics (tetracycline, flagyl), Lasix, antiarrythmics (amiodarone)	Detailed history, reviewing medications lists. Note if any new medications started or dosage changes.
Malignancy • Pancreatic or ampullary tumors	History: Constitutional symptoms Imaging: CT Abdo with contrast. *consider if recurrent AP NYD
Infectious • Viral: CMV, HIV, HBV, Cocksackie, varicella, HSV, • Bacterial: legionella, leptosporidium, salmonella • Fungal: Aspergillous • Parasites: Ascaris, Cryptosporidium	History: Ask about exposures and associated symptoms Laboratory: Depending on infectious agent suspected.
Autoimmune	History: Asking about any PmHx of autoimmune diseases. Laboratory: IgG (subclass 4), CT can be useful (retroperitoneal fibrosis) Treatment: Steroids
Scorpion Stings/Spider Bites • Only scorpions found in Caribbean	History: Ask about exposures and associated symptoms.
Hypercalcemia/Hyperparathyroidism	Extended lytes
Hypertriglyceridemia	History: Family history of hypertriglyceridemia, and acquired causes (Obesity, DMT2, Tamoxifen use, glucocorticoid use, nephrotic syndrome) Laboratory: Fasting lipid profile
Hereditary pancreatitis • PRSS 1, SPINK 1, CFTR	History: Ask about family history of pancreatitis.
Anatomical • Pancreas Divisum • Sphincter of Oddi Dysfunction	Imaging
Vascular • Vasculitis (SLE) • Athroembolic • Intraoperative hypotension	Imaging Lab: ANA screening for SLE

Complications of Acute Pancreatitis:

Local	Pancreatic	Nonpancreatic
	• Acute Peripancreatic Fluid Collection • Pancreatic pseudocyst • Acute Necrotic Collection (sterile or infected) • Walled off Necrosis (sterile or infected)	• Ileus • Gastric outlet dysfunction • Ascites or pleural effusion • Bile duct obstruction • Splenic vein thrombosis • Portal vein thrombosis
Organ Failure	**Cardiovascular** • SIRS/Shock • Hypotension secondary to capillary leak→third spacing→hypovolemia	**Renal** • Oliguria • Renal failure •
	Pulmonary • Hypoxemia • Atelectasis • Pleural effusion • ARDS	
Systemic	This is described as an exacerbation of any underlying disease (eg. Someone with underlying CAD going on to have an episode of demand ischemia)	
Other complications	**Metabolic** • Hyperglycemia • Hypocalcemia • Hypertriglyceridemia • Metabolic acidosis	**Hematologic** • DIC
	Hemorrhagic • Stress gastritis/ulcers • pseudoaneurysm	

Evaluation Severity:

Many scales have been created to assess the severity of pancreatitis include the Modified Glasgow Criteria of Severity, Bisap score, APACHE II and Ranson's Criteria. These scales help to differentiate the 75-85% of people who will have a mild course of pancreatitis to those who will have a complicated course and higher mortality. The AGA now recommends using the following criteria to stage the severity of pancreatitis:

- *Mild AP*→ No organ failure and no local complications and no systemic complications
- *Moderately Severe AP*→0-48h of organ failure and/or local complications and/or systemic complications
- *Severe AP*→SIRS on admission and/or Organ failure persisting for >48h and/or death (accounts for 15-20% of cases) **NB to define organ failure, must have 2 points on the Modified Marshall Scoring System*

Modified Marshall Scoring System:

Organ system	Score				
	0	1	2	3	4
Resp (PaO_2/FiO_2)	>400	301-400	201-300	101-200	<101
Creatinine(assuming a normal baseline)	<134	134-169	170-310	311-439	>439
Systolic BP	>90	<90, fluid responsive	<90 not fluid responsive	<90, pH<7.3	<90, pH<7.2

Management of Pancreatitis:

- General supportive care:
  - If severe pancreatitis, admit to the ICU
  - IV fluids: Early fluid resuscitation with RL
  - Pain management: opioids/antiemetic (IV ondansetron if QTc is normal)
  - Monitor ins/outs
  - Correction of any metabolic abnormalities
- Nutritional support:
  - Mild AP→may start feeding as soon as n/v/abdo pain resolves
  - Moderately severe→Nasojejunal tube feeding, using an elemental or semi-elemental formula, is preferred over total parenteral nutrition. Total parenteral nutrition should be used in those unable to tolerate enteral nutrition due to very severe AP or refractory ileus.
- Antibiotics
  - Give if they have any extrapancreatic infections (i.e. concomitant cholangitis, bacteremia, etc), but no empiric treatment needed for sterile necrosis.
  - If patient fails to improve or deteriorates after 7-10, consider infected necrosis and consult IR/sx for CT-guided biopsy OR may just use empiric treatment with carbapenems, quinolones or metronidazole.

Gallstone pancreatitis:

- Urgent ERCP (within 24 hours) should be performed in patients with gallstone pancreatitis who have concomitant cholangitis. Early ERCP (within 72 hours) should be performed in those with a high suspicion of a persistent common bile duct stone (visible common bile duct stone on noninvasive imaging, persistently dilated common bile duct, jaundice).
- ERCP, sphincterotomy, pancreatic duct stent alone provides adequate long-term therapy. In all others with gallbladder in situ, definitive surgical management (cholecystectomy) should be performed in the same hospital admission if possible and, otherwise, no later than 2–4 weeks after discharge.

ELEVATED LIVER ENZYMES

Diagnostic Approach: None of the tests is specific for liver disease but overall pattern and relative magnitude of abnormalities in liver enzymes often provides diagnostic clues to the type of liver disease.

Hepatocellular:

- **↑ AST&ALT**

>15x ULN	
DDx:	**Ix:**
Ischemic (shock liver)	Clinical picture, other parameters of hypoperfusion
Acute viral hepatitis • Hep A, B, C, D, E, CMV, HSV, EBC, VZV, schistosomiasis, toxoplasmosis	HAV IgM, HBsAg, HBcIgM, HCV Ab (if negative consider CMV, monospot, HSV Ab, VZV Ab)
Drugs/toxins	Salicylate level
Autoimmune hepatitis	ANA, anti-smooth muscle Ab, quantitative Ig, anti-LKM Ab, anti-SLA Ab
Budd-Chiari	Doppler U/S
HELLP	B-HCG
Rhabdo	CK
<15x ULN	
DDx:	**Ix:**
Viral hepatitis (chronic)	HBsAg, HBcIgM, HCV Ab
Fatty Liver	U/S
Drugs	Review med list
Autoimmune hepatitis (chronic)	ANA, anti-smooth muscle Ab, quantitative Ig, anti-LKM Ab, anti-SL Ab
EtOH	AST:ALT > 2:1
Wilsons	Ceruloplasmin
Alpha-1 antitrypsin deficiency	Alpha-1 antitrypsin level
Hemochromatosis	Serum transferrin saturation
Malignancy (liver, breast, small cell Ca, lymphoma, melanoma, myeloma)	Imaging as indicated
Cholestatic	Imaging as indicated
Other: acute fatty liver of pregnancy, glycogen storage dx, liver surgery, Reye's syndrome with viral illness and ASA use, legionella pneumonia, myopathy, strenuous exercise	See below

If AST and ALT are in the 1000s, consider ischemic hepatitis, acute auto-immune, and acute viral.

Common culprit drugs: acetaminophen, NSAIDs, amiodarone, statins, phenytoin, valproic acid, Isoniazid, Ketoconazol, Rifampin, tetracyclines, halogen anesthetics, mushroom, heavy metals, anabolic steroids, cocaine, ecstasy, phencyclidine

Cholestatic:
- **↑ ALP and Bili**
- **Order an abdo U/S looking for biliary dilatation**

U/S shows biliary dilatation = Extrahepatic Cholestasis	U/S absence of biliary dilatation = Intrahepatic cholestasis
CBD Stone PSC Stricture Pancreatic cancer Cholangiocarcinoma Send for ERCP	PBC (antimitochondrial Ab) PSC (Hx UC; Quantitative Ig, p-ANCA, anti-smooth muscle Ab, ANA, MRCP) EtOH Drugs TPN Sepsis Infiltrative Dx Consider MRCP

Infiltrative Pattern:
- **↑ Increase in ALP/GGT +/- bilirubin/AST/ALT**
- Infectious – TB, histoplasmosis, abscess (bacterial, amoebic)
- Neoplasm – hepatoma, lymphoma
- Granulomatous disease – sarcoidosis, TB, fungal
- Others – amyloidosis

Interpreting hepatitis serologies:
- HAV IgM = current HAV infection
- HBsAg +ve = current HBV infection
  - HBsAg and IgM anti-HBc (+HBeAg, HBV DNA, ↑ ALT) = acute infection
  - HBsAg and IgG anti-HBc (+anti-HBe, HBV DNA) = chronic infection
- Anti-HBs +ve = vaccinated, or resolved previous infection
  - Anti-HBs and IgG anti-HBc +ve (+/- anti-HBe)= resolved previous infection
  - Anti-HBs and IgG anti-HBc –ve = vaccinated
- Anti-HBe, HBV DNA, and ALT all indicate degree of infectivity
- HCV Ab screen +ve = send HCV RNA, if both +ve indicates infection

ACUTE LIVER FAILURE

Definition: development of hepatic encephalopathy and coagulopathy (INR ≥1.5) within 26 weeks of onset of illness.

Management:

1. Symptom Control – ABC, O2, IV hydration, check sugars routinely, monitor ICP
2. Coagulopathy – Vitamin K 5-10mg IV, FFP 2-4 U if active bleeding
3. AKI – supportive renal replacement. Consider midodrine, octreotide and albumin in Hepato-Renal Syndrome
4. Confusion – Look for signs of cerebral edema or Encephalopathy – Consider hyponatremia, infection, treat encephalopathy (see section that follows), intubate as necessary
5. Treat the underlying cause, consider NAC (even in non-acetaminophen ALF).
6. Consider liver transplant in patient with fulminant liver failure if no contraindication

DISEASES OF THE GALLBLADDER

Risk Factors: Female, Age, Fertile (Parity and pregnancy), Race (Indian>Hispanic>White), Obesity (BMI>30 or rapid weight loss), Pharmacology (Octreotide/Ceftriaxone), chronic hemolysis (pigment stones)

Biliary Colic:
- A gallstone (GS) transiently impacted in cystic duct; GB contracts secondary to CCK = transient RUQ/epigastric pain, worse after eating, fatty foods
- No infection. Normal WBC, liver enzymes
- Tx: elective lap chole

Acute Cholecystitis:
- GS impacted in cystic duct = GB inflammation
- Fever, anorexia, N/V, positive Murphys (97% sens, 48% spec), ↑ WBC, normal liver enzymes
- US (distended GB, wall>5mm edema, pericholic fluid, stone in CBD, sonographic Murphys). If U/S negative consider HIDA
- Tx: Abx (ie Cipro/flagyl), lap chole, preferably while in-patient

Choledocholithiasis:
- GS impacted in CBD (usually stone from GB, but consider biliary stasis secondary to PSC, CF, biliary stricture)
- Light stool, dark urine, jaundice, pruritus
- Normal WBC, ↑ liver enzymes/bili
- Tx: ERCP (consider MRCP if unsure of dx), Consider cholecystectomy if recurrent

Ascending Cholangitis:
- GS impacted in CBD → causes primary bacterial infection proximally from organisms that ascend from the duodenum (ie Klebsiella, E. coli, Enterococcus, Proteus, Pseudomonas)
- Charcot's triad: RUQ pain, fever, jaundice; Reynolds' pentad: triad + shock, confusion
- ↑ WBC, ↑ liver enzymes/bili
- Tx: ERCP within 24-48h, perc transhepati cholangiography if ERCP not available. Abx. Cholecystectomy once cholangitis resolves

HEPATIC ENCEPHALOPATHY

Definition: HE is a brain dysfunction caused by liver insufficiency and/or portosystemic shunting and manifests as a wide spectrum of neuro and psychiatric abnormalities ranging from subclinical alterations to coma

- Will occur in 30-40% of people with cirrhosis at some time during their clinical course

Stages of Hepatic Encephalopathy (West Haven Criteria)

	Grade	LOC	Orientation	Intellect/ Behaviour	Neurologic Findings
Covert	Minimal	Normal	Normal	Normal	Diagnose via a psychometrical/ neuropsych test
Covert	I	Mild lack of awareness	Change in sleep rhythm	Short attention span, impaired addition/ subtraction, euphoria/anxiety	Normal
Overt	II	Lethargic/ apathy	Disoriented to time (≥3 wrong: day of the month, day of the week, month, season, year)	Personality change (apathy, irritable, disinhibited), inappropriate behaviours	Asterixis (*NB:* asterixis=overt HE by definition)
Overt	III	Somnolent but arousable	Disoriented to space (≥3 wrong: country, province, city or place)	Gross disorientation; bizarre behaviour, confused	Muscular rigidity, clonus, hyperreflexia, upgoing babinski
Overt	IV	Coma		Coma	May be hypotonic and hyporeflexic

Precipitants of Hepatic Encephalopathy:

- Increased NH3 production – GI bleed, Infection (SBP, Hep C), transfusion, azotemia, hypokalemia, increased protein intake, constipation
- Increased diffusion across BBB – alkalosis
- Medications – benzos, narcotics, EtOH, med non-compliance
- Dehydration – diuretics, hemorrhage, large volume paracentesis, vomiting, diarrhea,
- Other – portosystemic shunt, progressive liver damage (HCC), portal vein thrombosis

Diagnostic Approach to Hepatic Encephalopathy:

1. **History**: confusion (onset, duration, fluctuation), sleep/wake reversal, neurological symptoms, precipitants, past medical history (liver disease, alcohol, illicit drug use), medication history. Collateral history very important.
2. **Physical Examination**: vitals, stigmata chronic liver disease, asterixis, rectal examination, neurological exam, Reitan Trail test (should finish in their age. i.e. 30yo=finish in 30sec)

3. **Investigations**:
    - Blood Tests – CBC, lytes, BUN, Cr, Glucose, TSH, Bili, AST, ALT, ALP, INR, PTT, Albumin
    - Microbiology – Blood C&S, Urinalysis, Urine C&S. NB-Low sensitivity and specificity with ammonia concentrations – not required for diagnosis of HE
    - Imaging – US abdo with liver doppler, (CT head), (CT abdo)
    - Special – ascitic fluid analysis to rule out SBP, gastroscopy if signs of UGI bleed

Treatment:

- Identification & treatment of precipitating factors – important to rule out SBP
- An episode of overt HE (whether spontaneous or precipitated) should be treated
    - 1st line: prevention of ammonia absorption from the bowel by altering stool acidity and changing NH3→NH4 – **Lactulose 20-30 g (30-45 ml) every 1-2 hours** to induce rapid laxation (goal=2 loose/soft BM), then titrate to achieve 2-3BM/d (usual daily dose: 60-100 g (90-150 ml) daily or Lactulose 200 g (300 ml) diluted with 700 mL of H_2O or NS, administer rectally via rectal balloon catheter and retain 30-60 minutes every 4-6 hrs)
    - 2nd line: reduction/elimination of substrates for the generation of nitrogenous compounds-achieved by use of antibiotics that reduce colonic bacteria (**Rifaximin 550 mg BID**) NB-Use Rifaximin with Lactulose, not alone
    - 3rd line: short-term restriction of dietary protein (1g/kg/day) – considered only in severe encephalopathy, long-term restriction cause worsening malnutrition
    - Consider doing brain CT to rule out other etiology if no improvement within 48 hours.
- Secondary prophylaxis with lactulose after an episode of OHE is recommended
    - If have recurrent bouts of OHE while on lactulose alone, add Rifaximin
- Primary prophylaxis for prevention of OHE is only required in patients with cirrhosis and a high risk of developing HE
- Intractable OHE with liver failure is an indication for a liver transplant assessment

MANAGEMENT OF ASCITES

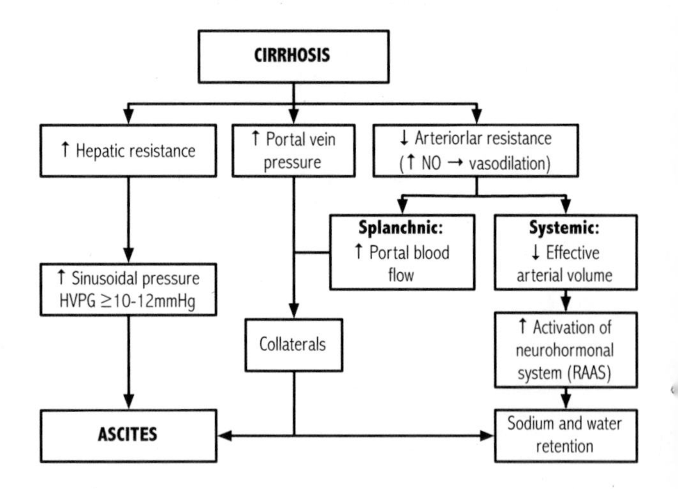

Determining the Source of Ascites (by Serum-Ascites Albumin Gradient (SAAG) and Ascites Protein)

Source of Ascites		
Most common causes: Cirrhosis 75% HCC 10% CHF 5% TB 2% Pancreatits 1%	**Portal Hypertension SAAG > 11**	**Ascites protein < 25** Sinusoidal hypertension (Cirrhosis)
		Ascites protein > 25 Post-sinusoidal hypertension: -Cardiac (CHF, constrictive pericarditis) -Budd-Chiari -Massive liver mets -Myxedema
	Peritoneum SAAG < 11	**Peritoneal lymph Ascites protein > 25** -Malignancy (ovarian cancer/peritoneal carcinomatosis), -Peritoneal TB, -Nephrotic syndrome -Pancreatitis (amylase >100)

Dx:

- Do a diagnostic ± therapeutic paracentesis
- Send the fluid for: Albumin, Total protein, Cell count and differential (to rule out SBP- see below), C&S, Cytology, Amylase
- Send blood for Albumin (To calculate SAAG)

Uncomplicated Ascites (i.e. no SBP, no HRS) Stages:

- Grade I-Mild→Only dx with U/S→ No treatment needed
- Grade II-Moderate→Moderate distension of the abdo (can detect on PE when >500cc in abdo)
- Grade III-Severe→Gross abdo distension→Large volume paracentesis (For every 3L taken off over 5L in the paracentesis give 1 bottle of albumin 25% in 100cc IV *need blood product consent)

Treatment for Moderate/Severe Ascites

1. Treat the cause of ascites
2. Stop EtOH consumption
3. Sodium Restriction<88mmol or 2g sodium per day
    - Note **NO** fluid restriction unless hyponatremia
4. Diuretic Therapy (to institute for Grade II/III):
    - Spironolactone and Lasix in 100:40mg ratio (these should be started and their dose increased simultaneously)
    - The dose should be adjusted every 3-4 days to a maximal effective dose of 400 mg/day spironolactone and 160mg/day lasix.
    - Target weight loss of 0.5kg/day max (without leg edema) and 1kg/day max (with leg edema) in order to prevent diuretic-induced renal failure and hyponatremia
    - Reduce dose once ascites is stabilized and follow

Diuretic complications:

- Renal failure→Typically due to intravascular volume depletion
- HE→ a RF for HE, but mechanism unknown
- Electrolyte disorders
    - Hypokalemia with Lasix and hyperkalemia with aldo antagonist
    - Fix lyte disorder before commencing diuretics
        - D/C Lasix if K<3
        - D/C Spironolactone if K>6
    - Hyponatremia→Stop diuretics if sodium drops to <120mmol/L
- Gynecomastia (with spironolactone)
    - Do not stop diuretic. Consider eplerenone.
- Muscle cramps
    - If very severe, stop or decrease the dose of diuretic and give albumin

Most of these complications occur during the first week, so careful monitoring of Cr and lytes during this time

1. Initial Large Volume Paracentesis (LVP) for tense ascites
2. Once diuretic refractory ascites, consider TIPS +/- liver transplant, continue to do LVP, continue diuretics only if urinary sodium excretion >30mmol/d

SPONTANEOUS BACTERIAL PERITONITIS (SBP)

Definition: An ascitic fluid infection without an evident intra-abdominal source

Presentation of SBP:
- Common presentation: fever, diffuse abdominal pain, jaundice
  - Tmax >37.8 as many cirrhotic patients are hypothermic
- Worsening encephalopathy
- Shock - hypotensive and hypothermia
- May be asymptomatic

Work up:
- Diagnostic paracentesis and send fluid for C&S, cell differential, cell count, albumin, total protein, glucose, LDH, amylase, bilirubin (if fluid is dark orange or brown) and cytology. Send serum blood cultures.
- Culture will only be +ve in 50% **therefore do the tap before giving antibiotics** (unless patient unstable)
  - inoculate blood culture bottles at bedside increases yield to 80%
  - Typical bacteria: *Klebsiella, E. coli, Enterococcus, Pseudomonas, Strep, Staph*

Diagnosis of SBP:
- Ascitic PMN count ≥ 250 (multiply the total nucleated cell count by the % PMNs)
  - Note, if fluid is hemorrhagic, 1 PMN is subtract from the PMN count for every 250 RBCs
- If the culture grows one species, but PMN<250…
  - If symptomatic (eg. fever)→ treat
  - If asymptomatic → repeat paracentesis in 48h and tx if PMN count rises to ≥250cells/mm^3

Differential Diagnosis:
- Secondary peritonitis from a perf, intra-abdo abscess, peritonitic TB or peritoneal carcinomatosis
- Think about a secondary cause if the gram stain shows many different kinds of bacteria
- Runyon's criteria can also help to discern between primary and secondary SBP
  - If have 2 of the following, think about secondary causes:
    - Total protein >10g/L
    - Glucose <2.8mmol/L
    - LDH the upper limit of normal for serum
  - Do a CT and abdo Xray to rule out another cause if these criteria are met
  - If nothing presents itself, Tx broadly and repeat paracentesis in 48h

Treatment of SBP:
- **Ceftriaxone 2g IV q24h x 5 days (if bacteremic extend to 2weeks)**
- If ? secondary causes add metronidazole
- Step down is Cipro 500mg PO BID x7d
- D/C nonselective beta blockers-higher risk of mortality and HRS

- Albumin (1.5g/kg initially, and 1g/kg on day 3)is indicated to help prevent renal failure in SBP if:
  - o BUN > 11mmol/L
  - o creatinine > 88umol/L
  - o bilirubin > 68umol/L
- Albumin is not indicated in patients with: community-acquired SBP, no GI hemorrhage, no encephalopathy, & normal renal function

SBP prophylaxis – 3 indications:
- History of SBP - prophylaxis with Norfloxacin 400mg PO Daily or Ciprofloxacin 500mg po daily
- Variceal bleeding – give Ceftriaxone 1g IV q24h x7d
- Cirrhosis and ascites if the ascitic fluid protein is <15 g/L along with either impaired renal function (Cr >106, BUN >8.9 or serum Na <130) or liver failure (Child-Pugh score ≥9 and a bilirubin ≥51 micromol/L)
  - o Treat the above with norfloxacin 400mg PO daily or ciprofloxacin 500mg PO daily

HEPATORENAL SYNDROME

Definition: The occurrence of renal failure in a patient with advanced liver disease in the absence of an identifiable cause of renal failure (thus making this a diagnosis of exclusion)

Presentation of HRS:
- A progressive rise in serum creatinine
- A normal urine sediment
- No or minimal proteuniuria (<500mg/d)/hematuria
- A very low rate of sodium excretion (<10meq)
- Oliguria (although not necessarily true early in the disease)

2 types of HRS

	HRS Type 1	HRS Type 2 (may progress to HRS 1)
Survival	1 month untreated	6 months untreated
Diagnosis	2x ↑ in Cr from baseline in <2 wks to >221umol/L	Cr >133 that slowly increases
Clinical	Typically occurs shortly after a precipitating factor	Associated with diuretic-resistant ascites

Precipitants:
- GI bleed
- SBP (30% of patients with SBP will get HRS→why give abx+albumin for SBP in some cases-see section on SBP)
- Large volume paracentesis
- NSAIDS
- Note: diuretics do not cause HRS

Diagnosis:

- A diagnosis of exclusion
- Proposed diagnostic criteria include:
  - Cirrhosis with ascites
  - A serum Cr >133umol/L that progresses over days to weeks
    - Note: A Serum Cr may be low as there is minimal Cr production in this malnourished population
  - An absence of renal parenchymal disease as identified by proteinuria <0.5g/d, no hematuria (<50rbc/hpf) and a normal renal U/S
  - No current/recent nephrotoxic drugs
  - No shock
  - No improvement in renal function (ie. Cr doesn't fall below 133) after volume expansion with IV albumin for at least 2 days and withdrawal of diuretics (thus ruling out hypovolemia)
    - Do 1g/kg/d up to a max of 100g/d: options for albumin are 25% in 100ml or 5% in 500ml

Treatment:

- Treat causes of liver failure
- D/C K-sparing diuretics (risk of hyperkalemia), lasix should be d/c'd unless need to treat central volume overload
- Avoid nephrotoxic drugs
- Monitor (vitals, urine output), if HRS type 1→ consider ICU, LFTs, Cr/BUN, lytes/extended lytes
- Screen for sepsis: BCx x2, urine R/M & C/S, ascitic fluid cultures
  - If no signs of infection, do not give antibiotics. Continue prophylactic antibiotics if previously given
- Ok to do a large volume paracentesis with albumin for tense ascites
- If non-ICU setting/Type 2
  - Midodrine starting dose 2.5-7.5mg PO q8h, then increase every 8h to a dose of 12.5mg q8h
  - Octreotide 100mcg SC q8h and increase to 200mcg SC TID, or as infusion 50mcg/h
  - Albumin→ initially a bolus of 1g/kg followed by 25-50g/d until midodrine and octreotide are d/c
- Early nephrology consultation

Treat for a total of 2 weeks. If no improvement in 2 weeks, then medical management is futile and should consider TIPS or a liver transplant with dialysis bridging to that.

DELIRIUM

Definition: acute change in attention and mental status with fluctuating course plus disorganized thinking OR altered level of consciousness.
Incidence: in patients > 65yo, 11-25% upon admission. Additional 30% may develop delirium.
Risk Factors: dementia (up to 50%), ICU admission (up to 70%), post-op (up to 50%) depression, visual impairment, EtOH use.
Prognosis: Associated with increased mortality. May persist for days-months (up to 21% still delirious months after initial Dx).
DDx: Dementia, depression, mania, acute psychosis, post-ictal state

Etiologies ("DIMS")	Examples
Drugs (intoxication or withdrawal)	Narcotics, anticholinergics (TCAs, phenothiazines, anti-Parkinsonians, GI and urologic drugs, Gravol, antihistamines, NSAIDs), benzodiazepines, EtOH, street drugs, steroids, nicotine
Infection	Most commonly UTI and pneumonia. Consider CNS infections, skin and soft tissue infections (check for pressure ulcers), endocarditis.
Metabolic	Hypoxemia (PE), hypercarbia (resp failure), liver failure, uremia, dehydration, B12 def, lytes (sodium, calcium, magnesium)
Structural	Trauma, tumour, stroke, vasculitis, ACS, CHF, urinary retention, constipation/fecal impaction, post-surgery
Other	Seizure, pain, starvation, new environment, visual/functional impairment, social isolation

Approach: History most important, use collateral; focus on DDx/triggers
- **History**: Time course? Association with other events? Pre-existing cognitive or sensory impairments (how were they 2 weeks ago)? Fluctuating course? Any possible alternate diagnosis? Visual/hearing aids? Hallucinations? Delusions?
- **Meds**: careful med review including PRNs, regular meds, OTC meds and EtOH. Reduce or eliminate as many possible culprit drugs
  - High risk meds: anticholinergics, benzos, narcotics
- **Physical**: vitals, inattention, asterixis, multifocal myoclonus, picking behaviour, hydration status, fecal impaction (do DRE), urinary retention, infected ulcers.
- **Investigations**:
  - Everyone: CBC, lytes, Cr, glucose, extended lytes.
  - If not done recently: TSH, Vit B12
  - If clinical suspicion: VBG/ABG, liver enzymes, blood cultures, urine R&M+C&S, CXR, ECG, trop, drug levels.
  - CT Head only if focal neuro signs or fall, EEG if suspect seizure, LP if infection/meningitis.
- Cause is usually **multifactorial**

Treatment:

- Treat underlying condition
- Supportive therapy – **minimize sensory deprivation** (glasses, hearing aids, frequent re-orientation); ensure good **hydration and nutrition**, **control pain** (start with acetaminophen), **increase mobility** (ambulation or up for meals) and **decrease tethers** (Foley, IV, restraints). Nicotine replacement. Clean up the medication list. **Delirium order set (at HHS)**.
- Monitoring: **CAM (delirium) score** regularly when checking vitals.
- Prevention: **Familiar faces/family visits/familiar objects** as often as possible, especially if agitated. Orient patient with **window, calendar**, and **clock** frequently (>3x/day). Have the patient in a **quiet** and well-lit room. Hospital Elder Life Program volunteers, if possible PSW at bedside to reorient and prevent falls.
- Insomnia: non-pharm measures (ear plugs, dark room); **melatonin** 3-6mg PO QHS
- Pharmacologic therapy (only if safety of patient or others is threatened, or pt in significant distress) – see below. Note antipsychotics may worsen parkinsonism.

Drug and dose	Adverse Effects	Additional Notes
Conventional anti-psychotic **HALOPERIDOL** **0.25–1.0 mg po BID; extra doses q4h prn** (peak effect, 4-6h) **0.25–1.0 IM**; observe & repeat in 30–60 min prn (peak effect: 20–40 min)	• EPS; especially if dose is >3 mg per day • Prolonged QTc • Avoid in patients w/withdrawal syndrome, hepatic insufficiency, • NMS	Usually agent of choice, especially if patient is violent. Suggestion of effectiveness demonstrated in RCTs Avoid IV use because of short duration of action & greater risk of QTc prolongation, arrhythmia Associated with increased mortality rate among older patients with dementia
Atypical anti-psychotics **RISPERIDONE** **0.25-0.5 mg twice daily** **OLANZAPINE** 1.25–5.0 mg once daily **QUETIAPINE** 6.25-25 mg twice daily	• EPS equivalent to low-dose haloperidol (<4.5mg) • Prolonged QTc on ECG (quetiapine likely greatest risk) • Quetiapine best for parkinsonism (less D2 antagonism)	Limited data from randomized controlled trials Likely non-inferior to conventional anti-psychotics (Cochrane review) Associated with increased mortality rate among older patients with dementia Quetiapine is sedating, so may be useful at nighttime.
Benzodiazepines **LORAZEPAM** 0.5–1.0 mg orally, with additional doses every 4 hr as needed	• Paradoxical excitation • Respiratory depression • Over-sedation • Ataxia • Falls	**Reserve for pts w/sedative and/or alcohol withdrawal, parkinsonism or NMS.** **DO NOT ROUTINELY USE FOR DELIRIUM.** Associated with prolongation & worsening of delirium symptoms (as demonstrated in clinical trials)

CODE WHITE

Acute treatment for patient behaviours that put themselves/others at risk:

- Ensure sufficient staff to restrain patient; do not put yourself or patient at risk
- **Haldol 0.25-1 mg IM q 30min PLUS lorazepam 0.5-1 mg IM q 30min** to max 4 doses until patient is no longer dangerous/distressed. Use the minimum effective dose, and be mindful it make take time to clear. Physical restraints must have permission from family, and may be associated with increased morbidity. Always use chemical restraints before physical restraints.

FALLS

Incidence of self reported injuries related to falls in the community: 57.5 per 1,000 age > 65yo

- 35% result in fracture
- Associated with significant morbidity and mortality
  - Decline in function, increased likelihood of LTC placement
  - Complications from falls are the **fifth leading cause of death** in adults > 65yo

Risk Factors: dementia (up to 50%), ICU admission (up to 70%), post-op (up to 50%) depression, visual impairment, EtOH use.

Intrinsic	Extrinsic
Acute illness	Behavioural
Balance and gait deficits	Fear of falling
Chronic conditions	EtOH
Neuro – Parkinson's, stroke	Environmental
Endo – DM2	Home environment
GU – bowel/bladder incontinence	Seasonal (higher in winter)
Psych - depression	Medications
Cognitive impairment	Antidepressants
Poor vision	Antipsychotics
Muscle weakness	Benzodiazepenes
	Antihypertensives

Approach:

- **History**: get best possible history surrounding falls from patient, caregivers, witnesses. Focus on circumstances surrounding the fall (what was the patient doing, what environment were they in). Screen for above risk factors for falls.
- **Meds**: get best possible medication history and ask about adherence/possible overdosing.
  - High risk meds: SSRIs, antipsychotics, benzos, sedative hypnotics, antihypertensives, alcohol
- **Physical exam:** orthostatic vitals (if able), neurological exam for a) any focal deficits or b) evidence of movement disorders. **Watch every patient's gait unless they are unable to walk!**
- **Investigations**:
  - As directed based on presence/suspicion of above risk factors
  - Do not order routine echo, Holter, radiographic investigations unless suggested by history/physical
  - CT head should be considered if:
    - Canadian CT head rules: GCS < 15 2h post injury, suspected fracture, sign of basilar skull fracture, > 2 episodes of vomiting, age > 65yo, dangerous mechanism of fall
    - Other: patient on therapeutic anticoagulation, focal neurological deficits on exam

Management:

- Treat underlying cause/risk factors for falls
- Early involvement of PT/OT
- Patients with recurrent falls (> 2 in last year) should be referred for a multifactorial falls assessment by a geriatrician as an outpatient

HEMATOLOGY & THROMBOSIS

ANEMIA

1. Is patient clinically stable? If no → resuscitate, transfuse as required
   **try to draw workup bloodwork before transfusion when possible*
2. Is anemia acute or chronic? Check previous Hb, note any hemodynamic instability; patient presenting with blood loss, angina/ischemic changes
3. Check mean corpuscular volume (MCV)

Microcytic (MCV <80)	**Normocytic** (MCV 80-100)	**Macrocytic** (MCV >100)
- Iron deficiency (↓ ferritin, ↑TIBC, ↓ TSAT) - Thalassemia - Anemia of chronic inflammation (normal/↑ferritin, ↓ TIBC, ↓ TSAT) - Sideroblastic anemia - Lead poisoning - Copper deficiency	- acute blood loss‡ - hemolysis (see below) - anemia of acute or chronic inflammation - sequestration (splenomegaly) - marrow infiltration/production problems‡‡ - chronic kidney disease (↓ epo) - endocrinopathy (eg, ↓ T4, Addison's) - pregnancy (physiologic) - early iron deficiency - early vitamin B12/folate deficiency or combined nutritional deficiencies	- reticulocytosis - vitamin B12/folate deficiency^ - liver disease - alcohol - hypothyroidism - myelodysplastic syndrome (MDS) - Drugs (methotrexate, hydroxyurea, AZT)

‡at least 15% blood volume loss before Hb change
‡‡ Look at other cell lines

^folate deficiency very rare in Canada due to supplementation, testing requires approval by Medical Biochemistry (RBC folate, homocysteine, MMA)

4. Determine bone marrow response. **Check reticulocyte count?**
   - If **ELEVATED** → consider acute blood loss, hemolysis, or recovering bone marrow (e.g. treated iron deficiency)
   - If **INAPPROPRIATELY LOW** → consider marrow disorder or marrow suppression
5. Look at blood film

Common blood film comments	Significance
RBCs	
Target cells	Hemoglobinopathies, liver disease, asplenia
Fragments/Schistocytes	MAHA (DIC, TTP, aHUS, HELLP)
Spherocytes	Hereditary spherocytosis, immune hemolysis
Howell-Jolly Bodies	Asplenia
Burr cells	Uremia
Teardrops	Myelofibrosis, marrow infiltration, hemoglobinopathy

Blister cells	Oxidative hemolysis
WBCs	
Left shift	CML, acute leukemia, marrow infiltration, infection
Toxic granulation	Infection
Hypersegmented neutrophils	Megaloblastic anemia
Myeloblasts	Acute leukemia if blast count >20%
Platelets	
Platelet clumping	Activation of platelets in EDTA, will result in false low count, send in citrated tube
Giant platelets	Peripheral platelet destruction (ex. ITP), myeloproliferative neoplasm (ex. ET, PV)

Anemia Clinical Pearls:

- Serial Hb useful: normal RBC destruction <0.5%/day, therefore a decrease of >10g/L/day (excluding lab error) Hb drop suggests blood loss or hemolysis; consider iatrogenic (laboratory testing).
- Therapeutic response: Hb rises ~10g/L/wk following repletion of iron/vitamin B12/folate with moderate reticulocytosis; continue supplementation until 3-6 months following normalization of Hb to replenish stores.
- Thalassemia trait vs. iron deficiency anemia (IDA): In **thalassemia** trait, MCV will be disproportionately low compared to Hb (eg, Hb 110, MCV 60), while RDW may be normal and RBC count elevated. In IDA, the MCV may be small but the RDW is usually high. TIBC is also increased in IDA, while ferritin is low . Interpret ferritin with caution in inflammatory states as it is an acute phase reactant and may mask underlying iron deficiency.
- Multiple lineages: If multiple cell lineages are reduced, consider a primary bone marrow disorder.

Hemolytic anemia:

- Evidence of red cell destruction (LDH/indirect bili↑, haptoglobin↓) and ↑red cell production (↑retic)
- Findings to confirm hemolysis: peripheral blood film (fragments, spherocytes, bite/blister cells, sickle cells), LDH (↑), haptoglobin (↓), free Hb (↑)
- Labs to determine etiology of hemolysis: Coombs test (direct antiglobulin test), cold agglutinin screen, G6PD/PK screen, coagulation parameters (DIC), flow cytometry/FLAER (PNH)

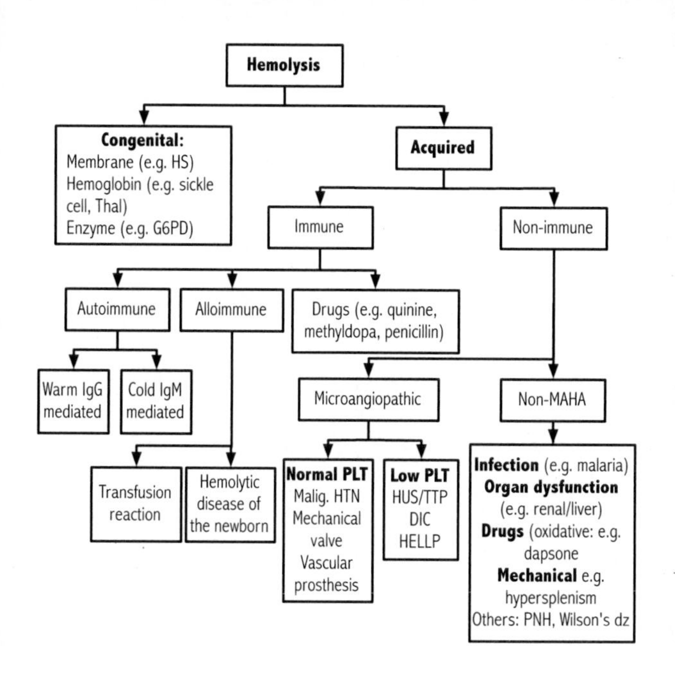

ANTICOAGULATION

Warfarin:

- Vitamin K antagonist with half-life approximately 40 hours.
- Anticoagulant effect is monitored using international normalized ratio (INR). Target range is INR of 2-3 except for mechanical mitral valves (2.5-3.5).
- Patients admitted to hospital may have an INR outside therapeutic range due to acute illness, change in diet (decreased vitamin K consumption) and medications (especially antibiotics). Increase monitoring until INR stable.
- **Starting dose**: 5 mg PO x 3 days; titrate to target INR (in patients <60 years, can start with 10 mg for 1-2 days followed by 5 mg dosing). Elderly patients or those on dialysis may require lower doses (1- 2 mg per day). INR measurement 3-4 days after first dose.

- **Monitoring frequency**: Measure INR 2-3 days after first dose. It takes approximately 3-7 days for full therapeutic effect. Once a stable INR is achieved the frequency of monitoring can be reduced (if stable; 1-2 times/month).
- Conduct a medication review when prescribing warfarin due to multiple drug interactions. Increase frequency of monitoring if changes to concurrent drugs (addition or removal)

Warfarin Reversal:

INR	Indication	Management
>1.5	Urgent surgery (< 24 hours)	Discontinue warfarin. Vitamin K 2.5 mg IV, Consider coagulation factor replacement with prothrombin complex concentrate (PCC; Octaplex® or Beriplex® **requires Thrombosis approval**)
>1.5	Urgent surgery (< 24 hours)	Discontinue warfarin. Vitamin K 1-2 mg PO (or IV if NPO). Repeat INR on morning of surgery.
Any	Major bleeding*	Discontinue warfarin. Vitamin K 5-10 mg IV and PCC (Octaplex®/Beriplex®) 20 units/kg for INR 2-3 30 units/kg for INR 3-6 40 units/kg for INR >6 Maximum dose 3000 units
5-9	Supratherapeuti c INR with no bleeding	Omit 1to 2 doses of warfarin. Reduce maintenance dose (decrease total weekly dose by 10%). Consider Vitamin K 1-2.5mg po for patients at increased risk for bleeding. Recheck INR in 24 hours.
>9	Supratherapeuti c INR with no bleeding	Omit 1 to 2 doses of warfarin. VitK 2.5-5mg PO. Recheck INR in 24 hours.

*Life, limb or organ-threatening bleeding. For all of the above scenarios, increase INR monitoring.

LMWH and UFH Reversal:

- UFH Reversal
  - Non-urgent: Stop infusion 4-6 hrs prior to procedure
  - Urgent: Stop infusion. Administer protamine sulfate 25-50mg over 15-30min. Protamine 1 mg reverses 100 U of UFH. Need monitored setting.
- **LMWH Reversal** – protamine reverses up to 100% anti-IIa activity but only 60% anti-Xa activity
  - Non-urgent: Check renal function. Last dose 24hrs prior to procedure. If high bleed risk procedure give 50% dose on the day before procedure. If impaired renal function discontinue 48 hours prior to procedure.
  - Urgent procedure or bleeding: Stop LMWH. If LMWH given in previous 8 hrs, 1mg protamine/100 anti-Xa units (Dalteparin, Tinzaparin). If LMWH given greater than 8 hrs, 0.5mg protamine/100 anti-Xa units (Dalteparin, Tinzaparin).

Direct Oral Anti Coagulants:

Dosing			
Stroke prevention in Non-valvular Afib	VTE prophylaxis after TKA/THA surgery	VTE treatment	Extended prevention of recurrent VTE
Dabigatran (Pradaxa) Direct Thrombin Inhibitor Not recommended for CrCl <30			
150mg PO BID 110mg PO BID if age>80	220mg PO daily	150mg PO BID (after a 5-10 days of parenteral anticoagulant)	
Rivaroxaban (Xarelto) Factor Xa Inhibitor Not recommended for CrCl <30			
20mg PO daily 15 mg PO daily if CrCl 30-49 mL/min	10mg PO daily	15mg PO BID x 21 days then 20mg PO daily	10 mg PO daily after 6 to 12 months of treatment
Apixaban (Eliquis) Factor Xa Inhibitor Not recommended for CrCl <25			
5mg PO BID 2.5mg PO BID if 2 out of the following: Age>80, Cr>133, weight<60kg	2.5mg PO BID	10mg PO BID x 7 days then 5mg PO BID	2.5mg PO BID following at least 6 months treatment
Edoxaban (Lixiana) Factor Xa Inhibitor Not recommended for CrCl <30			
60 mg PO daily 30 mg po daily if CrCl 30-50 mL/min		60 mg PO daily 30 mg po daily if CrCl 30-50 mL/min (after 5-10 days of parenteral anticoagulant)	

See Thrombosis Canada website www.thrombosiscanada.ca for further treatment and prophylaxis guidance. THA=total hip arthroplasty, TKA=total knee arthroplasty.

DOAC Reversal (within 18 hours of dose) for Major Bleeding or Urgent Surgery (within 8 hours):

- *Consult Order Sets at HHS or SJH*

1. Hold anticoagulant
2. Supportive care: monitored setting, volume replacement, blood product transfusion, refer for procedural/surgical intervention
3. Reversal
   - Dabigatran:
     - o Idarucizumab 5g IV (given as two 2.5g doses less than 15 minutes apart) **[Thrombosis approval]**
     - o Hemodialysis if practically feasible (~60% dialyzable)
     - o If idarucizumab not available consider activated PCC (FEIBA®) 2000 units IV **[Thrombosis approval]**
   - Rivaroxaban, Apixaban, Edoxaban: PCC (Octaplex®/Beriplex®) 2000 units IV **[Thrombosis approval]**

VENOUS THROMBOEMBOLISM

VTE Prophylaxis Considerations:

- Consider risk of VTE vs. bleeding for all hospitalized medical or surgical inpatients
- Choices include: dalteparin 5000 U subcut daily, enoxaparin 40mg subcut daily, tinzaparin 4500 U subcut daily, heparin 5000 U subcut BID. Avoid UFH when possible – increased risk of HIT. Fondaparinux 2.5mg subcut daily if history of HIT.
- Additional considerations: (1) Check patient's renal function (prophylactic dalteparin and tinzaparin can be used in renal dysfunction (above 20 mL/min). (2) Under- or overweight patients may require dose adjustment.
- When unable to use pharmacologic prophylaxis, consider intermittent pneumatic compression (IPCs, "moonboots") and encourage mobilization

Risk of Thromboembolism (Virchow's Triad):

- **Stasis:** immobilization, lymphatic/venous insufficiency, pregnancy, recent travel, plaster cast, etc.
- **Injury:** trauma, fracture, surgery (especially hip, leg, pelvic, gynecologic), central venous catheter
- **Hypercoagulable states:**
  - o *Acquired:* malignancy, lupus anticoagulant, thrombocytosis, pregnancy, puerperium, contraceptives and smoking
  - o *Inherited:* protein C & S deficiency, antithrombin III deficiency, prothrombin gene mutation and factor V Leiden

History:

- Query the risk factors.
- Unilateral limb swelling, pain, redness. Sudden onset of SOB, pleuritic chest pain, syncope.

Physical Examination:

- Unilateral leg (>3cm on affected side 10 cm below tibial tuberosity) or arm swelling, warmth, erythema, palpable cord, low grade fever.
- Homan's sign (pain in the calf on passive or active dorsiflexion) is *not* helpful as +LR=1 and –LR=1
- Tachycardia, tachypnea, hypoxemia. Massive PE – pulsus paradoxus, hypotension (sBP<90), RV heave

Well's Score for Deep Venous Thrombosis:

Criterion	Points
Active cancer (on-going treatment or within 6 months or palliative)	1 point
Paralysis, paresis, recent plaster cast of legs	1 point
Recently bedridden for ≥3 days or major surgery within last 4 weeks	1 point
Localized tenderness along deep venous system	1 point
Calf swelling ≥3 cm than asymptomatic leg (measured 10 cm below tibial tuberosity)	1 point
Pitting edema confined to symptomatic leg	1 point
Collateral non-varicose superficial veins	1 point
Alternative diagnosis at least as likely as DVT	-2 points
SCORE: High probability ≥ 3 (53%), Mod probability = 1-2 (17%), Low probability ≤ 0 (<5%)	

Low probability patients (Wells<2) and negative D-dimer rules out DVT without U/S. Moderate-high risk, negative D-dimer is not helpful.

Note: Wells criteria *not* validated for hospital inpatients – only use for outpatients.

Well's Score for PE:

Criterion	Points
S/Sx of DVT	3 points
Alt Dx less likely	3 points
HR >100	1.5 points
Immobilization or surgery in last 4 wks	1.5 points
Prior PE/DVT	1.5 points
Hemoptysis	1 point
Active Cancer	1 point
SCORE: High Probability ≥ 6 (70%), Mod = 2-6 (30%), Low <2 (<10%)	

*can also use PERC score to help rule out PE

Note: Wells criteria *not* validated for hospital inpatients – only use for outpatients.

Investigations:

- "classic EKG changes": sinus tachycardia, new atrial fibrillation, RVH, RAD, S1/Q3/T3.
- "classic CXR changes": Hampton's hump (pleural/pulmonary infarction), Westermark's sign (prominent pulmonary arteries proximal to PE and loss of peripheral vasculature distal to PE)
- **D-dimers** are not specific, but up to 97% sensitive. If low clinical pre-test probability and D-dimer is negative, D-Dimer helpful to RULE OUT VTE. However, if positive, then you must pursue further work up, so be careful in your utilization of this test. A negative D-dimer is not helpful in those cases where you have a moderate-high

suspicion of PE. **Recommendations suggest a D-Dimer NOT be done in pregnancy to help rule out DVT/PE.*

- **Doppler U/S:** lack of compressibility of a proximal vein in a patient without prior DVT is highly diagnostic of DVT.
- **Spiral CT (CTPA)** is the first line study. It may also provide an alternate diagnosis. V/Q scan is used when there is a contraindication to CTPA (renal failure, see pregnancy below
- **V/Q Scan** is helpful, but intrinsic lung abnormalities (COPD, asthma, CHF, other respiratory diseases) make interpretation difficult.
  - **If normal**: PE very unlikely. Do not treat.
  - **If indeterminate**: Evaluate bilateral lower extremities with compression ultrasounds. If pre-test probability is low, do not treat. If pre-test probability is high, do pulmonary angiogram.
  - **If high probability**: Diagnostic, therefore treat.
- **In Pregnancy:**
  - **See OB Medicine Section**

Treatment:

- *See anticoagulation/Length of treatment suggestions below.*
- If acute proximal DVT and anticoagulation is strictly contraindicated (i.e. bleeding, recent major surgery with high bleed risk), then consider temporary IVC filter until anticoagulation feasible (interventional radiology).
- If clinical pre-test probability for VTE is high, start treatment while awaiting imaging confirmation.
- If a patient with VTE is otherwise well, has low bleeding risk, and is reliable then consider discharging on therapy with Thrombo clinic follow-up.
  1. Give first dose of your treatment of choice (see below for details)
  2. Counsel patient about bleeding symptoms (urine, stool, muscle) and recurrent VTE symptoms. Document that you did this.

In Hospital PE Management:

1. ABCs. Oxygenate. Massive PE may require intubation and ventilation.

2. If hemodynamic instability (hypotension, sBP<90) call RACE/CCRT/CODE BLUE and consider systemic thrombolysis. Usual dose of alteplase (TPA) is 100 mg IV over 2 hours or 10 mg IV push followed by 90 mg over 2 hours. In the event of cardiac arrest, call CODE BLUE and follow ACLS protocol (50 mg IV push followed by additional 50 mg if no ROSC). If patient is unstable at outset, consider bedside ECHO to confirm diagnosis rather than sending the unstable patient to CT.
   - ***Absolute Contraindications*** include prior intracranial haemorrhage, ischemic stroke within last 3 months, aortic dissection, known AVM, known intracranial malignancy, recent neurosurgery, recent closed-head injury, active bleeding or bleeding diathesis
   - ***Relative Contraindications*** include age>75, internal bleeding in past 2-4wks, CPR>10min, severe uncontrolled hypertension, remove ischemic stroke (>3months), pregnancy, non-compressible vascular punctures, dementia, major surgery within 3 weeks.

3. The mainstay of treatment for intermediate risk PE (normotensive, but evidence of right heart strain on echocardiogram or CT and/or troponin rise) is prompt administration of therapeutic dose anticoagulation with close clinical monitoring. In the event of clinical deterioration, systemic thrombolysis can be considered in discussion with MRP and Thrombosis service.

Initial Anticoagulation for VTE:

1. DOAC
2. LMWH bridging to warfarin
   a. Start warfarin at same time as parenteral therapy (starting dose as above)
   b. Continue LMWH for a minimum 5 days, and INR must be therapeutic a minimum 2 days prior to discontinuing parenteral therapy
   c. Consider UFH for patients with severe renal impairment, patients with anticipated invasive procedure or if high risk of bleeding
3. LWMH only (e.g. cancer associated thrombosis, pregnancy)

Anticoagulant Dosing for VTE:

Drug	Dose
DOAC	Rivaroxaban 15mg BID x 21 days then 20mg daily Apixaban 10mg BID x 7 days then 5mg BID Dabigatran 150mg BID (after a 5-10 day initial treatment period with a parenteral anticoagulant)
LMWH	Dalteparin: 200 U/kg SC once daily or 100 U/kg SC twice daily. Enoxaparin: 1.5 mg/kg SC once daily or 1 mg/kg SC twice daily. Tinzaparin: 175 U/kg SC once daily.
UFH	Order set → weight-based IV nomogram (usually 80 U/kg bolus followed by 18-20 U/kg/hr infusion titrated to target aPTT (60-95 seconds) Subcutaneous regimen: 330 units/kg loading dose then 250 units/kg q12h
DOAC	Rivaroxaban 15mg BID x 21 days then 20mg daily Apixaban 10mg BID x 7 days then 5mg BID Dabigatran 150mg BID (after a 5-10 day initial treatment period with a parenteral anticoagulant)

Special Populations:

- **Cancer-associated VTE:** LMWH is therapy of choice
- **Heparin Induced Thrombocytopenia:** prior HIT < 100 days is an absolute contraindication to UFH/LMWH. In such cases, can use non-heparin anticoagulants (ex. Argatroban, Fondaparinux, Danaparoid). Acutely, warfarin should be stopped in the setting of HIT.
- **Pregnant patients**: Neither LMWH nor UFH cross the placenta (safe for the fetus). DOACs contraindicated in pregnancy. Warfarin avoided during pregnancy (risk of embryopathy)
- **Catheter-related DVT**
  - Goals of treatment: prevent recurrence, symptom relief, prolong catheter survival
  - Acute treatment similar to lower limb DVT
  - Do not need to remove catheter if still functioning, still required, and no evidence line infection.
  - Duration of anticoagulation: at least 3 months, or for duration of catheter use

Length of Treatment:

Condition	Duration of Anticoagulation
Transient **reversible** risk factor **(provoked)**	3 months if no ongoing risk factors
1st unprovoked	Minimum 3 months, then reassess based on risk factors for VTE recurrence + bleeding risk + patient preference
2nd unprovoked	Long-term anticoagulation for secondary VTE prevention, with periodic review
Cancer-associated VTE	Minimum 3-6 months Continue anticoagulation if patient is receiving systemic chemotherapy, has metastatic disease, has progressive/relapsed disease, or has other thrombosis risk factors

PERIOPERATIVE MANAGEMENT OF ANTICOAGULATION

For elective (non-emergent) procedures - Decision made considering the patient's risk for thromboembolic events and risk for bleeding in the perioperative period. The procedural bleeding risk determines the need for anticoagulant interruption and the timing of post-operative anticoagulant resumption. The thromboembolic risk determines the need for bridging anticoagulation (for patients on warfarin)

Bridging Anticoagulation for Patients Receiving Warfarin:

Bridging anticoagulation provides shorter-acting parenteral anticoagulant (e.g. low molecular weight heparin) to reduce the risk of thromboembolic events during the period of warfarin interruption. The need for bridging anticoagulation is determined by a patient's estimated risk for perioperative thromboembolism.

Risk of Thromboembolism:

Risk Category (% annual risk of thromboembolism)	Prosthetic Heart Valve	Atrial fibrillation	VTE
Very high risk (>10%) (Bridging anticoagulation recommended – consult Thrombosis)	• Any mechanical mitral valve • Caged ball/tilting disc aortic valve • recent (<6 mo) TIA or stroke	• Atrial fibrillation with $CHADS_2$** of 5-6 • recent (<3 mo) of TIA/stroke • rheumatic valvular heart disease	• recent (<3 mo) VTE • severe thrombo-phillia with history of VTE • prior thrombosis during anti-coagulation interruption
High risk (5-10%) (Consider bridging on case-by-case basis)	• Bileaflet mechanical AVR with or without risk factors for stroke*		• VTE in last 3-12 months • Recurrent VTE • Active cancer
Moderate risk (<5%) (Bridging generally not recommended)	• Bio-prosthetic heart valve	• Atrial fibrillation with $CHADS_2$** of 0-4	• prior VTE >12 mos ago

*Risk factors for stroke: Afib, HTN, prior stroke or TIA, DM, CHF, Age > 75
**$CHADS_2$ criteria: CHF (1 point); HTN (1 pt); age >75 years (1 pt); diabetes (1 pt); stroke/TIA (2 pts).

Risk of bleeding:

High Bleeding Risk	Moderate Bleeding Risk	Low Bleeding Risk
• Cardiac (CABG, valve replacement) • Neurosurgical (intracranial, intraspinal) • Vascular (AAA repair) • Urologic (prostatectomy, nephrectomy, kidney biopsy) • Extensive cancer surgery, vascular organ surgery • Any neuraxial procedure • Endoscopy with polypectomy	• Other abdominal surgery (hernia repair, laparoscopic cholecystectomy) • Endoscopy with bopsies • Lymph node biopsy	• EGD/sigmoidoscopy/ colonoscopy without biopsies • Cardiac catheterization/PCI • Dental • Cataracts • Skin • Common bedside procedures (thoracentesis, paracentesis, arthrocentesis)

Pre-operative Oral Anticoagulation Interruption for Elective Procedures/Surgery:

Drug	Minor Procedure (low bleeding risk)	Major Procedure (high bleeding risk)
Warfarin	Last dose 6 days prior to procedure, INR day before procedure with target INR ≤1.5. If INR above target can give vitamin K 1-2 mg po day before surgery.	Last dose 6 days prior to procedure, INR day before procedure with target INR ≤1.5 for most procedures. Target normal INR for some including neurosurgery, neuraxial anesthesia. If INR above target can give vitamin K 1-2 mg po day before surgery.
Dabigatran	If CrCl>50 mL/min: Last dose 2 days prior to procedure (skip 1-2 doses)	If CrCl>50 mL/min: Last dose 3 days prior to procedure (skip 3-4 doses)
	If CrCl 30-49 mL/min: Last dose 3 days prior to procedure (skip 3-4 doses)	If CrCl 30-49 mL/min: Last dose 5 days prior to procedure (skip 7-8 doses)
Rivaroxaban	Last dose 2 days prior to procedure (skip 1 dose)	Last dose 3 days prior to procedure (skip 2 doses)
Apixaban	Last dose 2 days prior to procedure (skip 2 doses)	Last dose 3 days prior to procedure (skip 4 doses)

Adapted from Thrombosis Canada www.thrombosiscanada.ca

Post-operative Management of Anticoagulation:

- For low bleed risk procedures, typically can resume therapeutic dose anticoagulation 24 hrs post-op
- For high risk procedures, delay restarting therapeutic dose anticoagulation until 48-72 hrs post-op. This decision should be guided by post-op hemostasis and risk factors for ongoing bleeding and discussion with surgeons.
- Prophylactic dose parenteral anticoagulation can be given 24 hours post-op for thromboprophylaxis if hemostasis is achieved.

BLOOD PRODUCTS AND TRANSFUSION REACTIONS

Informed consent is required for all blood component transfusions, including Albumin. If questions about possible transfusion reaction, can call Blood Bank. Bloody Easy available online at:
http://transfusionontario.org/en/documents/?cat=bloody_easy

Transfusion Guidelines/Blood Products:

Product	Guideines
Red blood cells 1 unit of pRBCs increases Hb ~ 10 g/L	• Transfuse 1 unit at a time and reassess unless significant bleeding • Patients with chronic anemia (eg sickle cell, iron deficiency) may not require transfusion, even for lower Hb levels. Transfuse only if symptomatic or other indications. • Consider if Hb < 70g/L for all patients, especially if acute • Consider if > 70 g/L but evidence of impaired tissue oxygen delivery (ie. pulmonary or cardiac dysfunction, symptoms of anemia, or active bleeding)
Platelets 1 adult unit of plts increases ~30 x 10^9/L	• Maintain >100 for neurosurgery • >50 for major surgery or significant bleeding • > 50-75 for epidurals and lumbar punctures • 20-50 for minor procedures • >20 for sepsis with fever • >10 for all other patients • For ITP, only transfuse for life-threatening bleeding
Plasma	Contains all coagulation factors. Usually 2-4 units (10-15 cc/kg) 1. INR > 1.8 and bleeding, or for anticipated procedure in patient with coagulopathy resulting from multiple coagulation factor deficiencies 2. Massive transfusion or microvascular bleeding and cannot wait 30-45 minutes for PTT/INR results 3. TTP
Cryoprecipitate	Contains Fibrinogen, vWF, FVIII, FXIII. Usually 10 units as one adult dose. 1. Microvascular bleeding (target fibrinogen level >1.0) 2. Massive transfusion (target fibrinogen level >1.5) 3. Post-partum hemorrhage (target fibrinogen level >2.0) 4. VWD or Hemophilia A with bleeding (if targeted therapies not available) 5. Factor XIII deficiency

Reactions:
For ALL transfusion reactions:

1. Interrupt transfusion and maintain IV access
2. Assess patient
3. Check patient and blood product identification
4. Notify blood bank and return blood if testing required
5. Initiate investigations and management as indicated

RED FLAGS: hypotension, tachycardia, chest/back pain, SOB, bleeding, rash >2/3 BSA, temp >39°C

Fever (1 in 300):

- Usually a febrile non-hemolytic transfusion reaction (FNHTR), but can be bacterial sepsis or acute hemolytic transfusion reaction
- Management:
  - If no red flags and no evidence of hemolysis, can continue transfusion slowly and give acetaminophen.
  - If red flags, consider haemolytic transfusion reaction or sepsis and initiate workup.
  - If significant and recurrent FNHTR, may pre-medicate with acetaminophen, or steroids. Antihistamines not effective.

Acute Hemolytic Reaction (1 in 40,000):

- ABO incompatibility is potentially fatal! 50% due to administering properly labelled blood to wrong patient. Can also be caused by other group incompatibilities aside from ABO (usually from prior alloimmunization)
- Watch for fever, chills, hemoglobinuria and other red flag symptoms.
- Result: DIC, shock, and acute renal failure due to acute tubular necrosis. Secondary to complement activation, intravascular hemolysis of transfused blood due to recipient ABO IgM antibodies.
- Management:
  - Give I.V NS (100-200ml/hr) to promote diuresis and maintain BP – consider diuretics if BP stable and need to maintain urine output > 100 mL/hour for ATN
  - Send to blood bank: donor's blood (for retyping and Coomb's Test), residual of blood product(s) and clamped tubing
  - Send to lab: Stat CBC & diff, INR, PTT, ext lytes, BUN, creatinine, urine for free Hb, cultures from blood product(s) and patient as well as group and screen
  - Manage DIC and hemorrhage as clinically indicated

Delayed Hemolytic Reaction (1:7000):

- 3-14 days following a transfusion, characterized by mild anemia and/or hyperbilirubinemia, positive DAT
- No specific Tx, unless red flags
  - Delayed hemolytic reaction characterised by extravascular hemolysis of previously transfused blood product due to recipient antibodies (IgG) against red cell antigens (Kidd, Duffy, etc.) acquired from previous transfusion/pregnancy. DHTR seen after re-exposure to blood containing minor antigen.
  - Blood Bank will inform patient of the name of the offending antigen. Advise patient to get medic-alert. Ensure appropriate hospital and blood bank records.

Urticaria (1 in 100):

- If no red flag symptoms, give diphenhydramine 25-50 mg IV/PO and restart slowly.
- *If previous reaction:* may prevent by pre-medicating with diphenhydramine and steroids, plasma depleting blood product or washed blood product.
- If any red flag symptoms, do not restart transfusion! Suspect an anaphylactic or severe allergic reaction, or acute haemolytic reaction

Anaphylaxis (1 in 40,000):

- Rapid onset of anaphylaxis, manifested by shock, hypotension, angioedema, and respiratory distress
- Majority of cases unexplained. Can occur in congenitally IgA deficient patients who mount an immune response to donor IgA.
- Consider acute hemolytic reaction (send off hemolytic workup). See "**Anaphylaxis**" section for treatment

Massive Transfusion: >10 units pRBCs in 24 hours. HHS has a protocol/order set ("Code Omega").

- Identify coagulopathic patients (eg. hemophilia) and initiate treatment (see below)
  - early use of tranexamic acid
- Use a blood warmer; adjust FFP, cryoprecipitate, platelet transfusion as per bloodwork
- Check CBC, lytes, PTT, INR q2H or q5pRBCs as point of care; replace as needed
  - Correct complications: dilutional coagulopathy, hypocalcemia, hyperK, hypothermia.

Transfusion-related Acute Lung Injury (TRALI) (1 in 10,000):

- Clinically diagnosed: acute onset hypoxemia within 6 hours post transfusion, dyspnea, bilateral lung infiltrates, PaO2/FiO2 ratio <300. Fever, tachycardia, and tachypnea are common clinical signs, and hypotension may occur.
  - Consider other transfusion reaction and send off appropriate blood work. May be confused with volume overload but JVP not elevated.
- Stop transfusion. Diuretics and steroids not helpful. Mechanical ventilation if indicated. Usually resolves in 24-72 hours. Supportive care for management.
- Blood Bank must be informed as TRALI is often due to antibodies in the donor's blood.

Transfusion-related Circulatory overload (TACO): (1 in 100)

- Clinical presentation of SOB, orthopnea, cyanosis, elevated JVP
- Consider diuretics with transfusion if history of CHF/volume overload. Limit transfusion volume and perform frequent clinical re-assessment if additional units are required.

Infectious Transmission

- Incidence per transfusion (Bloody easy 4 2016)
- Bacterial sepsis: 1:250 000 symptomatic , 1:500,000 death (1:200 000 for platelets)
- HIV: 1: 21 million
- HCV: 1: 13 million
- HBV: 1: 7.5 million
- Other infection transmitted by blood: West Nile virus, malaria, CMV, HTLV 1, parvo

COAGULOPATHIES

Prothrombin Time (PT or INR): measures factors II, V, X, VII, fibrinogen (extrinsic/common pathway)Isolated ↑ **INR:** vitamin K antagonists (e.g. warfarin), hepatic dysfunction,,vitamin K deficiency, DIC, factor VII inhibitor/deficiency

Activated Partial Thromboplastin Time (aPTT) measures factors II, V, X, VIII, IX, XI, XII, fibrinogen (intrinsic /common pathway). Isolated ↑**PTT** : heparin,dabigatran, congenital factor deficiency (e.g hemophilia A, B, C), acquired factor deficiency (e.g. hepatic dysfunction), specific coagulation factor inhibitor (e.g. acquired hemophilia with inhibitors for factors VIII or IX), antiphospholipid antibody (if reagents sensitive to non-specific inhibitor/lupus anticoagulant), VWD (secondary to factor VIII deficiency)

Increased aPTT and INR: Multiple coagulation factor deficiencies (DIC, liver disease, dilutional coagulopathy from massive transfusion), excessive anticoagulation, deficiencies/inhibitor in common pathway factors (X, V, II, fibrinogen)

Thrombin Time (TCT) measures conversion of fibrinogen to fibrin; 2 units Thrombin added to patient plasma.

- Prolonged TCT: heparin, hypofibrinogenemia, dysfibrinogenemia, direct thrombin inhibitors (e.g. dabigatran, argatroban).
- Heparin contamination can be detected by correction of TCT by adding 10 units of thrombin or protamine sulphate to patient sample

D-Dimer is a breakdown product of fibrin. D-dimer can be increased in setting of thrombosis, fibrinolysis (ie. DIC), inflammation, malignancy, or surgery.

History: Consider using bleeding risk tool (MCMDM-1 Bleeding Questionnaire)

1. Clinical features:
   - Active, acute/chronic bleeding?
   - One site or many sites? Gingival or traumatic (eg IV)?
   - Family history?
   - Drug history? (including OTCs and herbals)
   - Prior hemostatic challenges (surgeries; delivery; tooth extractions)
   - Prior transfusions

2. Pattern of bleeding:
   - **Primary** hemostasis (platelets, Von Willebrand's Disease): Mucocutaneous (eg gums); bruising, capillary oozing, petechiae, purpura, epistaxis, GU/ GI bleeding
   - **Secondary** hemostasis (clotting factor): (soft-tissue) hemarthrosis, hematomas, extensive bleeding, retroperitoneal bleeding.
   - NOTE: Intracerebral bleeding can occur with both primary and secondary hemostasis disorders.

Investigations: CBC and blood film, renal function, liver function, INR, PTT, TCT, Fibrinogen, consider coagulation factor level assay depending on result of PTT, INR.

Primary Hemostasis:

- Thrombocytopenia (platelet < 150,000/ul): **Refer to thrombocytopenia.**
- Platelet Dysfunction (normal INR, PTT and platelet count; consider sending platelet function analysis and aggregation studies):
  - **Inherited:** Bernard-Soulier, Glanzmann's thromboasthenia, storage pool disease
  - **Acquired:** Drugs (aspirin, clopidrogrel), Renal Failure, Liver Failure, myeloproliferative disorders (acquired VWD), DIC, myelodysplastic syndrome (acquired platelet function abnormalities).
    - Note: Bleeding Time no longer available; platelet aggregation studies if platelet function disorder suspected
- Von Willebrand's Disease: vWF:RCo, vWF:Ag, FVIII studies, VWF multimers

Secondary Hemostasis:

- Synthesized in Liver: I, II, V, VII, IX, X, XI, XII (factor VIII synthesised by endothelium)
  - Factor VII: shortest half-life (5-7hrs) therefore deficient 1st in early liver failure, DIC, Vit. K deficiency and warfarin therapy. INR first to be elevated.
- Vitamin K dependent: II, VII, IX, X
  - Vitamin K deficiency – Caused by: Fat malabsorption including biliary obstruction, malnutrition, pancreatic insufficiency, recent antibiotic use, warfarin use, liver disease- decreased storage of vitamin K)
  - Vitamin K antagonists (eg. warfarin)
- Hemophilia A & B: inherited factor VIII and IX deficiency, respectively
- Specific Factor Inhibitors:
  - Most commonly seen in hereditary hemophilia after factor replacement.
  - Acquired hemophilia: idiopathic (40%), malignancy, lymphoproliferative disorders, post-partum, autoimmune disease, *factor VIII inhibitor is most common, but can affect any factor.

COAGULOPATHY MANAGEMENT GUIDELINES

Treat the patient, not just the numbers! Call SMR +/- MRP Staff +/- hematologist on call for complicated problems.
*Refer to Transfusion reaction for transfusion product contents

Thrombocytopenia: Outside of chemotherapy-induced thrombocytopenia
- Approach: decreased production, increased destruction, or sequestration?
- Platelet transfusions are rarely indicated, unless patient is seriously bleeding or requires surgery.
- Treat underlying cause (eg. remove offending drugs if drug-induced). **Caution** in suspected TTP/HIT as platelet transfusion increases thrombosis risk.

ITP:
- If plt <20-30 and no bleeding: prednisone 1mg/kg daily (typically for 2-3 weeks followed by taper) or dexamethasone 40 mg po daily for 4 days
- If bleeding and/or severe thrombocytopenia (platelet count less than 10): IVIG 1g/kg daily for 1-2 days and prednisone 1mg/kg/day or dexamethasone 40 mg po daily for 4 days

Platelet dysfunction:
- Uremia - try DDAVP 0.3 mcg/kg IV/SC (max 20 mcg) and monitor for hyponatremia; other options include cryoprecipitate or estrogen therapy
- **Von Willebrand Disease:**
  - **Type 1, Type 2A**: trial of DDAVP 0.3ug/Kg IV – check vWF and FVIII levels afterwards to assess response, which usually persists for 6- 12 hours (done electively in a non-urgent outpatient setting). **[Call Hematology in the event of bleeding or surgery]**
  - **Type 2B, 2M, 2N, 3**: Humate-P or Wilate (WWF concentrate) for severe bleeding or prior to surgery **[Call Hematology]**

DIC:
- Involve Hematology +/- Critical Care.
- **Treat underlying cause!** (i.e. Sepsis, trauma, malignancy, obstetric complications, meds).
- If bleeding, support with plasma and cryoprecipitate (keep fibrinogen >1 gm/L).
- If clotting or purpura fulminans, call hematologist before starting heparin.

Hemophilias (hemophilia A: VIII, hemophilia B: IX):
- Replace with purified/recombinant factor VIII, or IX concentrate. DDAVP for mild disease (consider prophylactic for procedures). **[Call Hematology]**

Factor Inhibitors:
- Recombinant factor VIIa or activated PCC (FEIBA®), porcine factor concentrates, activated prothrombin complex, plasmapheresis, high purity human factors. **[Call Hematology]**

THROMBOCYTOPENIA

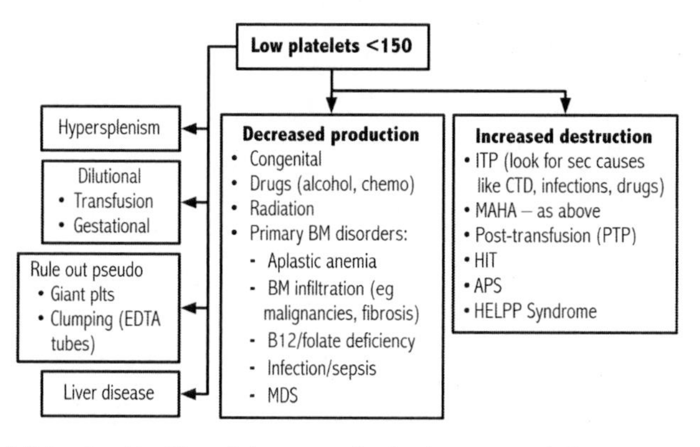

Initial workup: blood film, reticulocyte count, liver function tests, coagulation tests

Targeted workup: Medication review, vitamin B12 level, viral screen (HBV, HCV, EBV, HIV), autoimmune screen (if symptoms/signs suggestive of autoimmune disorder); consider HIT assay if heparin exposure, abdominal ultrasound, bone marrow biopsy (if other unexplained cytopenias or abnormalities on peripheral blood film)

Plts < 10 puts the patient at risk for spontaneous intracranial bleeding. Safety of antithrombotic medication should be evaluated when platelets less than 50.

HEPARIN-INDUCED THROMBOCYTOPENIA AND THROMBOSIS (HITT)

Autoantibodies against heparin and platelet factor 4 (PF4) complex leading to thrombocytopenia, thrombin generation and thrombosis

Risk Factors:
- UFH > LMWH
- F>M
- Full-dose anti-coagulation compared with prophylactic doses
- Surgical>Medical patients

Clinical manifestations and consequences:
- Causes both venous and arterial thrombosis
  - Venous: DVT/PE, warfarin-induced venous limb gangrene, cerebral vein thrombosis, adrenal hemorrhagic infarction (rare)
  - Arterial: Aortic/iliofemoral thrombosis and limb ischemia; MI, CVA

- May also have skin necrosis at heparin injection sites
- Occasionally may present acutely with transfusion reaction (respiratory distress, tachycardia, fevers, chills) immediately after heparin bolus

Diagnosis:
- **Thrombocytopenia**
  - Usually nadir in range of 30-100 x 109/L; however, drop in count of 50% with overall normal platelet count is also significant. Post-op expect platelet rise as acute phase reactant (unless on infusion).
  - Typically, platelet begins to drop 5-10 days after heparin exposure (70% of patients)
  - May occur immediately after exposure in the context of recent exposure to heparin (usually within last 30 days, may be up to 100 days)
- **Presence of HIT antibodies**
  - HIT Assay (ELISA reported in optical density): OD >1.0 strongly suggestive of HIT
  - Functional assay (serotonin release assay) is used as a confirmatory test
- **Presence of thrombus**
  - Either new, recurrent or progressive

4T Score for HITT

Points	2	1	0
Thrombocytopenia	>50% fall in plt count with nadir ≥20x10^9/L	30-50% fall in plt count OR nadir 10-19x10^9/L	<30% fall in plt count OR nadir <10x10^9/L
Timing of onset of platelet fall	Days 5-10; OR ≤ day 1 with heparin within last 30 days	>10 day or timing unclear; OR <1 day with heparin in prev. 31-100 days	<4 day (no recent heparin)
Thrombosis	New *proven* thrombosis; OR Skin necrosis; OR Acute systemic reaction with IV bolus of UFH	Progressive or recurrent thrombosis; OR Erythematous skin lesions; OR Suspected thrombosis	None
Alternative cause of ↓plt	None identified	Possible	Definite

Pretest probability: 6-8 points = high; 4-5 = intermediate; 0-3 = low
If low, consider other causes and do not send HIT Assay. If intermediate or high, send assay, manage as HIT until result available.
***Adapted from Warkentin TE, Heddle NM. Laboratory diagnosis of immune heparin-induced thrombocytopenia. Curr Hematol Rep. 2003; 2: 148-157.*

Treatment
If clinically suspicious of HIT
- Stop heparin from all sources (lines, flushes, dialysis)
- Send HIT assay

- Start anticoagulation with non-heparinoid anticoagulant (i.e. Fondaparinaux, lepirudin, argatroban, danaparoid) at **therapeutic** dose
- Consider leg dopplers to r/o DVT if no thrombosis already diagnosed
- **NO platelet transfusions** (can worsen thrombosis) **unless active bleeding (rare)**

If HIT assay positive:
- No thrombosis - anticoagulation with non-heparanoid anticoagulant (e.g. fondaparinux, danaparoid, argatroban) at therapeutic doses for 1 month. Can transition to warfarin once platelet count has normalized.
- Thrombosis - non-heparinoid anticoagulant and transition to warfarin when platelets have normalized.
  - Duration of anticoagulation at least 3 months
  - Warfarin: Should not be used acutely in HIT. Minimum of 5 days of overlap between warfarin and non-heparinoid anticoagulant. Stop non-heparinoid only when therapeutic INR for at least 2 days.
  - Oral factor Xa inhibitors (rivaroxaban, apixaban) may be considered in consultation with Thrombosis Service.

THROMBOTIC THROMBOCYTOPENIC PURPURA (TTP) / HEMOLYTIC UREMIC SYNDROME (HUS)

Medical emergency! Without prompt recognition and treatment, >90% mortality, <25% with Tx. Idiopathic TTP etiology generally a deficiency of ADAMTS13 – needed to cleave vWF – when uncleaved, platelets are hyperactivated leading to platelet consumption/ thrombus formation, anemia.

Clinical Features:
- Pts presenting with unexplained ↓in both plts and Hb MUST have a peripheral smear assessing for schistocytes (RBC fragments) to demonstrate microangiopathic hemolytic anemia (MAHA).
- If patient presents with MAHA and thrombocytopenia initiate treatment for TTP while awaiting further workup.
- CNS abnormalities (focal or nonfocal), renal abnormalities (proteinuria, hematuria, ARF) and fever (rare; high grade + chills => Sepsis>TTP-HUS) make up the rest of the "classic pentad" – only ~5% have full pentad at presentation.
- TTP and atypical HUS are similar; TTP generally worse thrombocytopenia, while atypical HUS has more renal impairment.

Differential Diagnoses (other causes of low platelets and microangiopathic hemolytic anemia)
- Preeclampsia/eclampsia/HELLP
- Sepsis, DIC (↑ INR/PTT & Hx/Px)
- Malignant hypertension
- Catastrophic Antiphospholipid Antibody Syndrome
- Vasculitis (SLE, scleroderma)
- Disseminated malignancy
- Heparin-induced thrombocytopenia
- Scleroderma renal crisis

History:
- Recent infection, HIV risk factors, bruising, rash, joint pains
- New medications: ticlopidine, clopidogrel, mitomycin, cisplatin, cyclosporine, tacrolimus, quinine (also in tonic water)
- Exposure to E. coli O157:H7 (meat, milk, cheese, animals) or Shigella; history of infectious diarrhea (risk factors for HUS)
- Recent transplant, total body irradiation, chemotherapy; gyne and pregnancy history
- Symptoms: bleeding; headache, confusion, seizures and other neurological symptoms. Nausea, abdominal pain.

Physical Exam:
- Fever, CNS focal signs; mental status, stroke
- Skin/oral petechiae, bruising
- Evidence of a secondary cause: infection, autoimmune disease, malignancy

Investigations:
- Thrombocytopenia, blood film with schistocytes (dense RBC fragments),
- Evidence of hemolysis: high reticulocyte count, negative Coombs Test, High LDH & unconjugated bilirubin, absent haptoglobin (sensitive but can be reduced in liver disease), free hemoglobin
- Creatinine, troponin
- INR and PTT (normal in TTP, abnormal in DIC)
- Urinalysis for evidence of hemoglobinuria (blood but no RBCs), BUN, ß-HCG, ECG, CXR, CT head (if CNS Sx), stool culture/O+P if bloody diarrhea, ANA, ANCAs, antiphospholipid antibodies

Therapy for TTP:
1. Call hematology to initiate THERAPEUTIC PLASMA EXCHANGE (plasmapheresis; PLEX)
2. If PLEX cannot be started immediately, give plasma infusion @ 100ml/hr (goal 25 to 30 mL/kg per day, limited by Vol overload) until PLEX/instead of PLEX
   - Replace 1-1.5x plasma volume daily with plasma or Cryo-poor plasma
   - PLEX: removes unusually large vWF that activates platelets, replaces ADAMTS13, removes antibodies to ADAMTS13
   - PLEX daily until platelet count recovers then tapered
   - Before initiating PLEX or plasma infusion, draw a coagulation tube and send to Platelet Immunology Lab (see McMaster Platelet Immunology Lab website for requisition) for ADAMTS13 assay (most cases of TTP have ADAMTS13 activity $< 10\%$)
3. Adjunctive therapy with corticosteroid (prednisone 1mg/kg)
4. Avoid platelet transfusions due to potential for thrombosis (only for bleeding or major procedure). Central venous catheter insertion for PLEX should be conducted by experienced individual.
5. Follow CBC, reticulocyte count, blood film (schistocytes), LDH daily
6. Refractory and relapsing courses: Rituximab, immunosuppression

Note: for suspected cases of atypical HUS, definitive treatment is Eculizumab.

SEPSIS

A spectrum of clinical symptoms of presumed infectious origin that includes:

- **Fever (T ≥ 38° C) or hypothermia (T < 36° C)** – in some (elderly, uremia, steroid use, cancer/immunocompromised), febrile response may be blunted.
- **Leukocytosis > 12,000/uL or < 4,000 or >10% bands**
- **Heart rate > 90 BPM** – increased cardiac output from peripheral vasodilation, due to endotoxemia and inflammation. Gives way to hypotension that is increasingly resistant to fluid challenge.
- **Respiratory rate > 20/min** - tachypnea and increased A-a gradient are signs of early ARDS or non-cardiogenic pulmonary edema. Initial respiratory alkalosis followed by combined acidosis.

Systemic inflammatory response syndrome (SIRS) = 2 or more of the above
Sepsis = SIRS + suspected infection
Severe sepsis = sepsis + organ dysfunction, hypotension, or hypoperfusion (eg. DIC, coagulopathy, renal failure, lactic acidosis, CNS changes)
Septic shock = severe sepsis + hypotension unresponsive to standard fluid bolus (10 mL/kg NS or RL). State of peripheral hypoperfusion and tissue ischemia.

Investigations:

- Basic bloodwork including: CBC, lytes, extended lytes, Cr, BUN, VBG/ABG, lactate (follow clearance)
- Microbiology: blood culture x 2 sets (helps identify whether culture is true or contaminant), culture from all lines, urine culture, sputum culture, stool if suspecting C.diff/infection
- Imaging: CXR, further imaging if concerned about abscesses
- Approach patient systematically to rule out potential sources of infection.

Management:

- See *ICU Section* **for complete management and target goals in sepsis. Start broad spectrum antibiotics until source can be narrowed and cultures return.**

NOTE: in conditions where there is no SIRS criteria, delay of antibiotics may be appropriate for full diagnostic workup and selection of antimicrobial therapy, as well as management of other conditions contributing to clinical presentation (eg. delirium).

Bacterial Meningitis:

Organisms:

- *S. pneumoniae* (GPC pairs/chains), *N. meningitidis* (GN diplococci), *H. influenzae* is less common due to vaccination.
- *Listeria monocytogenes* (GP bacilli) in young/elderly or immunocompromised (eg. steroids, transplantation, lymphoid and solid organ malignancies, chemotherapy, advanced HIV, pregnancy).
- Rarely tuberculosis (epidemiologic link, severely immunocompromised/HIV, previous TB).

Diagnosis:

- All patients with suspected meningitis should have two sets of blood cultures and lumbar puncture (LP) performed, unless contraindicated.
- **Send CSF for**
  - **Tube 1: cell count and differential, glucose and protein**
  - **Tube 2: gram stain and culture**
  - **Tube 3: virology - HSV PCR and other suspected organisms (ie. TB)**
  - **Tube 4: cell count and differential**
  - **Consider freezing extra CSF in tube 4 for further testing.**
- At our hospitals, all patients should be able to get a CT head prior to LP. You ***must*** get a CT scan before LP if immunocompromised (HIV, history of transplant, receiving immunosuppressive therapy), history of CNS disease, new onset seizure, papilledema, abnormal LOC or focal neurologic deficit.

All patients with meningitis require droplet and contact precautions until 24 hours after effective antibiotic therapy or another diagnosis is made. Notify Infectious Disease of all meningitis, especially nosocomial (empiric regimen does not apply).

CSF characteristics	Bacterial meningitis	Viral meningitis
Opening pressure	↑, >180mm H_2O	N or slightly ↑
WBC count	↑, >1000/mm³, PMNs	↑ (10-2000/mm³), often lymph
CSF/serum glucose	<0.3	Normal (>0.6)
Protein	Elevated (often >1g/L)	Normal or ↑
Gram stain	Organisms (~50-80%)	No organisms

Treatment: Never delay antibiotics! If a CT scan is indicated, draw blood cultures and administer antibiotics before the CT scan.

- Age under 50: **ceftriaxone 2g IV q12h** (give this first!) + **vancomycin IV 15-20 mg/kg over 1 to 2 hours q8-12h**.
- Age over 50: need to cover Listeria, add **ampicillin 2g IV q4h** or **Septra 5mg/kg IV q6h** if allergic to ampicillin.
- Steroids: Start **dexamethasone 0.15 mg/kg IV q6h for 2-4 days** if suspicious for *S. pneumoniae* **prior** to first dose of antibiotic.
- Treatment duration 7-21 days depending on organism.

Other Considerations:

- *Previous Neurosurgical procedure/ CSF shunt?*
  - o Start vancomycin 15mg/kg q8-12 hours depending on weight + ceftazidime 2g IV q8h *or* meropenem 2g IV q8h to cover *S. pneumoniae/S. aureus/Coagulase negative staphylococcus*
  - o Neurosurgical or Interventional Radiology consultation for sampling of CSF
- *Brain abscess risk?*
  - o Add metronidazole 7.5 mg/kg IV q 6h to cover anaerobes
  - o Neurosurgical consultation for aspiration of brain abscess, avoid sampling of CSF
- *Allergy to beta-lactams?*
  - o Vancomycin + moxifloxacin 400mg IV daily +/- Septra 5mg/kg IV q6-12h for *Listeria*
  - o Consider skin testing if during daytime hours, if overnight consult ID as may still benefit from low cross reactive beta lactams
- *Severely Immunocompromised?* (Solid organ transplant, hematologic malignancy, stem cell transplant, HIV with CD4 less than 100)
  - o Ampicillin + vancomycin + meropenem 2g IV q12h (broader GN coverage)
  - o Consult ID as patient may also need antifungal therapy

Viral Meningitis and Encephalitis:

- Common organisms: HSV, Varicella zoster, CMV, EBV, HHS6, Enterovirus, influenza, acute HIV, west nile virus, measles, mumps
- **HSV Encephalitis Treatment: Acyclovir 10mg/kg IV q8H for 21 days**

Assessment:

- Always look for source of bacteremia
  - urine, intra-abdominal abscess, chest, endocarditis, often skin and soft tissue including ulcers and wounds are missed
- Do not insert lines or prosthetic materials while patient is bacteremic (exclusions apply).

Organisms:

- Gram Positive
  - ***Staphylococcus Aureus (*mandatory ID consultation)***: Staph Aureus is essentially ***never*** a contaminant! Repeat blood cultures q24-48 hours for clearance of bacteremia.
    - All patients require a TTE at minimum, some may require TEE.
    - Treatment requires **minimum** 2 weeks of antibiotics.
    - Start **Vancomycin 15mg/kg q8-12 hours plus Cefazolin 2g IV q8h or Cloxacillin 2g IV q4h**. Step down antibiotics pending MRSA vs. MSSA and once cultures return.
    - Potential sources of Infection include: Central Lines, IV drug use, acute infective endocarditis, septic joints, osteomyelitis, pulmonary infection, skin and soft tissue infection, rarely meningitis/brain abscess
  - *Coagulase Negative Staphylococci (CONS)*: consider contaminant if only single source of blood culture or long incubation period. Look for source from lines or prosthetic materials (joints, drains, shunts).
    - Start vancomycin empirically if patient is sick, should do a clinical assessment before starting therapy.
  - *Enterococcus species*: **cephalosporins DO NOT work for enterococcus**; consider vancomycin, ampicillin, or ciprofloxacin for urinary sources
- Gram Negative
  - Common Organisms: E. coli, Klebsiella, Proteus, Enterobacter, Pseudomonas
  - Empiric therapy: consider ceftriaxone, ceftazidime or piperacillin-tazobactam (if concerned about pseudomonas), fluoroquinolones. **Once patient stable, afebrile x 48 hours, step down to oral agent to finish treatment course of 7-14 days.**
  - If there is a history of ESBL or patient is high risk (multiple prior antibiotics), consider meropenem or ertapenem empirically.

LINE INFECTIONS

Assessment:

- A line infection is a bloodstream infection evidenced by bacteremia/fungemia in the presence of an indwelling temporary (PICC, IJ, Subclavian, Femoral), or permanent (permcath, port-a-cath) line, in which the line is highly suspected to be the source of infection (or another source cannot be found).
- Cultures drawn from lines are often positive ≥2 hours before peripheral cultures drawn at the same time
- Draw blood cultures concurrently from line and peripheral source if possible
- May not have signs of localized infection but should examine each line site, if localized swelling, redness, could also consider Doppler ultrasounds as thrombosis/septic thrombophlebitis can present concurrently with infection

Management:

- Strongly consider **discontinuing/removing central line** if *Staphylocccus aureus, candida, pseudomonas,* resistant gram negatives, thrombosis, localized abscess or in septic shock/unresponsive to therapy
- Empiric therapy: Vancomycin 15mg/kg q8-12 hours +/- ceftazidime 2g IV q8h. Consider carpabenem, antifungal if patient severely ill. Step down based on organism.
- Uncomplicated line infections can be treated for 7-14 days after negative blood culture
- Minimum two weeks with *Staphylococcus aureus*

INFECTIVE ENDOCARDITIS

Assessment:

- Modified Duke criteria (2 major; 1 major + 3 minor or 5 minor)
- **Major**
  - Typical organism in two blood cultures (*S. viridans*, *S. bovis*, HACEK group, *S. aureus,* enterococcus) OR two positive blood cultures >12 hours apart, or at least 3 of 4 separate blood cultures
  - ECHO findings: evidence of endocardial involvement manifested by presence of vegetation/valvular abscess on echocardiogram OR presence of new regurgitant murmur
- **Minor**
  - IVDU or predisposing heart condition
  - fever >38° C
  - vascular phenomenon (arterial emboli, ICH, conjunctival haemorrhage, janeway lesion)
  - immune phenomenon (GN, Osler nodes, Roth spots, positive RF)
  - positive blood culture not meeting major criteria, echo abnormal but no vegetation.

Management:

- Tailor therapy to culture results. Obtain 3 sets of blood cultures initially, followed by every 1-2 days until cultures negative.
- Obtain ECG looking for conduction abnormalities (i.e prolonged PR) if aortic vegetation.
- Consult ID for advice around ongoing management and echocardiography. **All patients require a minimum of a transthoracic echocardiogram (TTE)**. Prosthetic valves, high risk individuals or high clinical suspicion for endocarditis require transesophageal echocardiogram (TEE).
- Empiric therapy: **Vancomycin 15mg/kg q8-12h** (depending on weight/renal function) + **ceftriaxone 2g IV daily**.
- Treatment duration for minimum 6 weeks, narrow antibiotics once culture speciation and sensitivity returns.
- If patient is clinically well and does not meet SIRS criteria, can consider delaying antibiotics to obtain cultures

Other Considerations:

- Prosthetic valve endocarditis will require addition of rifampin +/- gentamicin
- Cardiac surgery consultations may be required for valve replacement/repairment. Clearer indications for left-sided endocarditis than right-sided (ex. Ongoing embolic phenomena, refractory heart failure, refractory bacteremia, worsening heart block, size of vegetation)
- Complications of septic emboli (ie. lungs, brain) or heart failure requires monitoring
- Implanted cardiac devices (ie. pacemakers, ICDs) will require tailored treatment with possible consideration for replacement (consult cardiology first).
- Prophylaxis for dental procedures in persons with prosthetic heart valves, prior IE, with Amoxicillin 2g PO x1 prior to procedure
- HACEK organisms grow in routine blood cultures, ask lab to hold for 7 days
- Occasional forms of culture negative endocarditis – Bartonella, Q fever, Brucella, Legionella, Whipple's Disease

Community Acquired Pneumonia:

- Healthy outpatient:
  - Often unknown bugs. Consider *S. pneumoniae*, atypicals (*Legionella, Mycoplasma, Chlamydia*), viral
  - Azithromycin 500 mg PO x1 then 250 mg PO daily x5 days OR 5-7 days of amoxicillin 500mg tid
  - Can also use levofloxacin 750 mg/moxifloxacin 400mg once daily x5 days. Avoid quinolones for healthy outpatient pneumonia if possible (unnecessarily broad, increases resistance). Check QTC and renally dose
- **Comorbidities** (chronic heart, lung, liver, renal disease; T2DM; immunosuppressed; antimicrobials in previous 3 months) as **outpatient**:
  - Usually *S. pneumoniae*, gram negatives and anaerobes (alcoholics), *H. flu* or *M. catarrhalis* (COPD)
  - β-lactam (amoxicillin 1 g tid OR amoxicillin-clavulanic acid 875mg PO bid OR cefuroxime 500 mg po bid) PLUS macrolide (azithromycin as above OR doxycycline 100 mg PO bid)
- **Inpatient non-ICU**:
  - Consider *S. pneumoniae, H. influenzae, M. catarrhalis* and atypical organisms
  - β-lactam (ceftriaxone 1-2 g IV q24h) + macrolide (azithromycin as above) OR Respiratory FQ (levofloxacin 500-750 mg po/IV once daily or moxifloxacin 400 mg po/IV once daily) x5-7 days
- **Inpatient ICU**:
  - Non-pseudomonas: β-lactam (ceftriaxone) PLUS azithromycin OR respiratory floroquinolone
  - Pseudomonas: Anti-pneumococcal, anti-pseudomonal β-lactam (tazocin/imipenem/meropenem/ceftazidime) PLUS respiratory floroquinolone

Nosocomical/Nursing Home/Hospital Acquired Pneumonia(HAP):

- Requires broad GP and GN coverage.
- Treatment duration of 7 days. Consider:
  - Ceftriaxone 2g IV q24h; or
  - Levofloxacin 500-750 mg PO/IV once daily (HHS) or moxifloxacin 400 mg PO/IV (SJH); or
  - Piperacillin-tazobactam 4.5 g IV q6h if severe illness and/or high risk for *Pseudomonas aeruginosa*

Ventilator Associated Pneumonia(VAP):

- Requires broad GP and GN coverage, including for *Pseudomonas* and *S. Aureus*
- Send Sputum/Endotracheal tube aspirate sample for culture
- Treatment duration of 7-10 days. Consider:
  - Vancomycin for gram-positive and MRSA **and**
  - Piptazo, ceftazidime, or imipenem/peropenem for gram-negative and *Pseudomonas* **and**
  - Respiratory fluoroquinolone (cipro/levo) or aminoglycoside with pseudomonal activity for double coverage

Aspiration Pneumonia:

- Most macro-aspirations don't result in pneumonia
- Changes in oxygenation or CXR immediately following an aspiration event are usually "chemical pneumonitis" and don't require antibiotics
- If suspected aspiration pneumonia (eg. alcoholism, very poor dentition), cover usual oral anaerobes such as *Bacteroides, Peptostreptococcus*
- Consider using ceftriaxone, or amoxicillin-clavulanic acid for 5-7 days.

Influenza:

- Consider in patients with fever, headache, myalgia, malaise and respiratory tract symptoms if it is flu season and virus has been active in the community
- Send **nasopharyngeal swab (NPS) to confirm and initiate droplet precautions**
- May give Oseltamavir (Tamiflu) 75 mg PO bid x 5 days IF symptom duration < 48 hours (otherwise patient is considered outside of the window to benefit, consider therapy after 48hrs in only in medically ill patients)
- If CrCL 10-30mL/min, dose reduce to Tamiflu 75mg PO daily x 5 days.

URINARY TRACT INFECTIONS

Assessment:

- Colonization:
  - Do not treat **asymptomatic bacteriuria** unless patient is **pregnant** or before a **urologic procedure when bleeding is anticipated** (eg. TURP) or **renal transplant** patient within 3-6 months of transplantation
- Infection:
  - Defined by bacteriuria + pyuria + symptoms.
  - Lower UTI (cystitis) - no constitutional symptoms, only localized symptoms.
  - Upper UTI (pyelonephritis) in young: fever/chills, flank pain, nausea, vomiting
  - Upper UTI (pyelonephritis) in elderly: falls, confusion, aLOC, may not have fever, this is a diagnosis of EXCLUSION, and rule out any other causes of decompensation

Management:

- **Uncomplicated UTI:**
  - Usually E. coli, Proteus, Klebsiella, enteric GNs, S. saprophyticus, Enterococcus (*S. Aureus* in urine is rare, and usually indicates systemic bacteremia, switch foley and repeat Cx)
  - Septra DS 1 tab PO bid x 3d, nitrofurantoin 100 mg po bid x 5d, clavulin 875 mg PO daily x 3 days, if refractory or resistant ciprofloxacin 250/500mg PO bid x 3d
  - In men, consider above regimens excluding nitrofurantoin, and consider treating for 5-7 days. **Exclude urethritis and prostatitis in men**
- **Enterococcus:**
  - 2nd most common bug if over age 65 and/or nursing home.
  - Look for GP cocci on gram stain.
  - Uncomplicated and sensitive – amoxicillin 500 mg PO TID x 5-7 days
  - Complicated: Ampicillin 1-2g IV q8h, if resistant consider nitrofurantoin, ciprofloxacin if sensitive, or Vancomycin if complicated
- **Pyelonephritis:**
  - Outpatient – ciprofloxacin 500mg po bid x 7d, Septra DS 1 tab BID x14 d
  - Inpatient – Ceftriaxone 1-2g IV q24h, Cipro 400mg IV q24h. Consider broadening coverage if severe (eg. Piptazo 3.375g IV q6h)
  - Consider renal imaging if concerning for septic stone or obstruction requiring urological evaluation
- **Complicated UTI:**
  - Include: pregnancy, comorbidities, hospital-acquired, obstructive (ie. stone), prosthetic materials (ie. catheters, stents, tubes), anatomical abnormalities, transplants, immunocompromised patients, sepsis, male
  - Ciprofloxacin 500mg po bid x 5-7d/IV 400mg bid, IV ceftriaxone 1-2g q24h, carbapenems and narrow once sensitivities return. If febrile send blood cultures to rule out urosepsis.
- **ESBL (*E. Coli or Klebsiella*) or SPICE (*Serratia, Providencia, Proteus vulgaris, Citrobacter, Enterobacter*):**
  - Uncomplicated – Use Ciprofloxacin, Nitrofurantoin, or Septra if sensitive, if resistant can consider Fosfomycin 3g PO x 1
  - Complicated – Empiric – ertapenem 2g IV q24hours (if not pseudomonas), can step down to ciprofloxacin if sensitive

Intra-abdominal infections:

- Usually need definitive surgical care or abscess drainage for source control, but patients with intraabdominal sepsis require immediate antibiotic coverage
- Cover enteric organisms and gut anaerobes.
- **Cephalosporin or ciprofloxacin PLUS flagyl (500mg PO/IV bid), OR Piperacillin-Tazobactam OR Carbapenem if severe infection**
- Consider **antifungal** therapy if critically ill or not responding to therapy with reasonable source control

Acute Infectious Diarrhea:

- Perform volume assessment for dehydration and rule out acute abdomen on exam. Assess quality of diarrhea (bloody, watery), associated symptoms (vomiting, fever, tenesmus), sexual contacts, medications, sick contacts, and risk factors (travel, antibiotic exposure, dietary history, employment)
- Consider testing with the following conditions:
  - o Community acquired or traveler's diarrhea: Stool for C&S (*Salmonella, Shigella, Campylobacter, E.coli* O157:H7, *C. difficile*), virology
  - o Nosocominal diarrhea (onset >3d in hospital): stool for *C. diff* toxin
  - o Persistent diarrhea >7d: O&P – *Giardia, Cryptosporidium, Cyclospora*, +/- inflammatory screen
- Consider treatment if severe or immunocompromised – Ciprofloxacin, Ceftriaxone, Flagyl

Clostridium Difficile Infection (CDI):

- Suspect in any patient with risk factors (recent hospitalization, recent antibiotic use, PPIs), increasing frequency of diarrhea, watery stools, and rising WBC count.
- Send stool for C. diff toxin, rehydrate, place in contact precautions
- Consider empiric treatment if very sick or high probability until PCR result confirms diagnosis.
- Do not re-test for to test for cure because it may stay positive after treatment.

Category	Treatment
Mild/moderate: WBC<15, Cr stable	Metronidazole 500 mg PO TID x 14 days.
Severe CDI: WBC>15, Cr>1.5x baseline	Vancomycin 125 mg po QID x 10-14 days
Severe CDI with complications: hypotension or shock, ileus, or toxic megacolon	Flagyl 500 mg IV TID PLUS vancomycin 500 mg PO/NG QID If ileus, consider vancomycin PR
Recurrence	Start Vancomycin 125-250mg PO QID x 14 days and taper; Involve ID for taper and for consideration of fecal transplant

Non-purulent SSTI (e.g Cellulitis, Erysipelas):

- Usually Group A *Streptococcus*, or *S. aureus*, even in diabetics
- Often a diagnosis problem (eg. mistaking stasis dermatitis for cellulitis)
- Cultures (unless febrile) and superficial skin wabs not routinely recommended.
- **Mild infection**: Cephalexin 500 mg po QID, Clindamycin 600 mg po tid, Cloxacillin 500 mg po q6h, Amoxi-clav 875 mg PO BID. Treat for 5-7 days
- **Moderate infection**: Cefazolin IV 1-2g q8h, Ceftriaxone 2g IV q24h. Step down to PO antibiotics once improving and treat for 7-14 days.
- **Severe infection (signs of systemic infection)**: rule out necrotizing infection (see below). Treat broadly with vancomycin + piptazo.
- MRSA risk factors:
  - purulent drainage/abscess formation, high risk groups (history of IVDU, incarceration, military, MSM, athletes, high local prevalence of MRSA, previous colonization with MRSA, family history of MRSA or recurrent boils/abscesses).
  - Mild infections – consider Septra, doxycycline, clindamycin (check sensitivities), Linezolid (potential drug drug interactions)
  - Moderate to Severe Infections – Vancomycin or Daptomycin

Purulent SSTI (e.g Furuncle, Carbuncle, Abscesses):

- All of these need I&Ds to be sent for culture and sensitivity
- For mild to moderate infections, empirically treat with Septra or doxycycline, conside Keflex or Clavulin if MSSA
- For severe infections (failure of I&D and oral antibiotics, or signs of systemic illness), treat with vancomycin empirically. Step down therapy once cultures return

Necrotizing Fasciitis:

Assessment:

- Suspect in patients with pain disproportional to clinical findings, hard/wooden feeling of subcutaneous tissue, systemic toxicity, edema/tenderness beyond cutaneous erythema, crepitus, bullous lesions, or skin necrosis.
- **IT IS A SURGICAL DIAGNOSIS.** If necrotizing fasciitis is a possibility, consult surgery immediately.
- Type 1: (after abdominal/perineal surgery, trauma; seen in diabetics/PVD) – polymicrobial with anaerobes
- Type 2: (usually in the extremities) – monomicrobial with skin flora often in healthy patients

Management:

- Consult surgery (orthopaedics for lower limbs, plastic surgery for upper limbs, general surgery for abdominal, urology for scrotal/perinanal) for early/urgent surgical debridement
- ID consultation for help with empiric antibiotics
- All patients require Droplet and Contact precautions until 24 hours after effective antimicrobial therapy or another diagnosis is made
- Antibiotics for polymicrobial infection should gram positives including MRSA, gram negatives, and anaerobes

- Vancomycin/daptomycin/linezolid (for MRSA) PLUS
- Piperacillin-tazobactam or carbapenem or ceftriaxone plus metronidazole or fluoroquinolone plus metronidazole
- AND clindamycin for Group A strep. Often difficult to tell if type I or II. Can discontinue if no Group A strep isolated.
- Consider adding IVIG if consideration for group A strep and end organ failure/toxic shock

Diabetic Foot Infections:

Assessment:

- Organisms:
  - Mild infections usually typical skin organism (*S. Aureus and Streptococcus spp)*
  - Moderate/severe infections usually polymicrobial with gram negative and gram positive aerobes, as well as anaerobes (*bacteroides, clostridium, peptostreptococcus*)
- Rule out underlying osteomyelitis (see below section on treatment)
- Peripheral arterial assessment (ABI) to ensure appropriate delivery of antibiotic; consult vascular surgery if suspecting of ischemic limb, severe infection with potential need for amputation to control sepsis

Management:

- Obtain initial CRP if suspecting OM, blood cultures if febrile, appropriate consultations for definitive surgical treatment if failed medical therapy
- Mild-moderate infections: aerobic gram positive coverage with Ancef, Keflex, Clindamycin, or Clavulin
- Severe infection or chronic: braod spectrum to cover gram positive, gram negatives and anaerobes with Ceftraxone + Flagyl or Ancef + Ciprofloxacin + Flagyl or Piptazo or Carbapenems
- Consider adding Vancomycin or Septra for MRSA
- Consider Pseudomonas coverage with Tazo, Cipro, Mero
- Other Considerations:
  - Wound care consult for pressure offloading, dressings, debridement
  - Treat underlying diabetes, sugar control
  - Not all ulcers are infected, and not all ulcers require full antibiotic duration prior to healing

OSTEOMYELITIS

Assessment:

- Organisms
  - Typical skin organisms including *Staph and Strep species*, can be polymicrobial with gram negative organisms and anaerobes in severe infections
  - Consider in IV drug users (*S. aureus, P. aeruginosa, Serratia*), diabetics (aerobic + anaerobic GPC/GNR), sickle cell (*nontyphoidal salmonellae*), post-operative (*Staph*, GNB), animal and human bites (*eikenella, pasteurella*), nail through sneaker (*pseudomonas*)
  - Route of infection is hematogenous usually in central or long bones or contiguous from soft tissue infections
- Probe to bone, exposed or visible bone or prosthesis, or pus in bone is diagnostic of OM, don't need anymore testing
- Deep >3mm or large >2 cm^2 ulcers increases likelihood of underlying OM
- Image with Xray, MRI or CT scan, Bone Scan/WBC Scan (lower sensitivity), Bone Biopsy is gold standard however rarely done unless vertebral osteomyelitis
- CRP at the start of treatment to monitor progress
- Blood Cultures if febrile

Management:

- Underlying prosthetic material will need to be removed! For prosthetic joint infections consider consulting Infectious Disease and Orthopedics for management
- Ensure antibiotic route is absorbed (ie. good peripheral blood supply), and appropriate weight-based dosing
- Antibiotic Choice:
  - MSSA: Ancef 2g IV q6-8H or Ceftriaxone 2g IV q24H
  - MRSA: Vancomycin 15mg-20/kg IV q12H, check troughs and target levels of 15-20
  - Polymicrobial: Cipro/Levo + Flagyl, Tazocin, Carbapenem (if severe)
  - Pseudomonas: Ceftazidime, Cipro, Tazo, Meropenem
  - Narrow to cultures if applicable; consider PO step down for Abx with good bioavailability (ie. Fluoroquinolones and Flagyl)
- Duration: Minimum 6 weeks for acute, 12 weeks for chronic, follow CRP, remove prosthetic materials
- Vertebral osteomyelitis – Obtain blood cultures, and empirically treat if meets SIRS criteria. If stable, avoid treatment unless blood cultures are positive. If cultures are negative HIGHLY consider bone biopsy for culture

SEPTIC ARTHRITIS/BURSITIS

Assessment:

- Send needle aspiration for cell count, crystals, gram stain and culture.
- Orthopedics must be called for all septic joints for consideration of surgical therapeutic aspiration/ joint irrigation.

Category	WBC/mm3	Color	Viscosity
Normal	<150	Colorless/straw	High
Non-inflammatory	<3000	Straw/yellow	High
Inflammatory	>3000	Yellow	Low
Septic	>50,000	Pus	Mixed

Management:

- At risk for STDs?
  - Usually *N. gonorrhoeae*. Ceftriaxione or cefotaxime 1g IV daily q24H
- Not at risk for STDs?
  - Consider *S. aureus, Strep species* and GN bacilli
  - Start ceftriaxone 2g IV + Vancomycin 1g IV q12H – step down therapy once gram stain results available
- Prosthetic joint?
  - Call ID and Orthopedics
  - Joint aspiration and debridement should usually occur before antibiotics
  - If patient too unstable, can start ceftriaxone + vanco

FEBRILE NEUTROPENIA

Definition:
- Temperature >38°C, neutrophils <0.5 x 10^9 neutrophils/L
- Neutropenia is defined as an ANC <0.5
- Classify patients into high risk for serious complications vs low risk:
  - High risk: neutropenia expected to last >7 days, comorbid medical conditions including hemodynamic instability, mental status changes, GI symptoms, oral/GI mucositis, intravascular catheter infection, new pulmonary infiltrate, hypoxia, underlying chronic lung disease, uncontrolled or progressive cancer, hepatic/renal insufficiency
  - Low risk: neutropenia expected to last <7 days with none of the above risk factors

Assessment:
- Search thoroughly for site of infection: skin, head and neck including ears, sinuses and oropharynx, respirator, GI, all catheter sites, IV sites and lines.
- **Avoid DRE!**
- Pan culture patient from all sites and lines.

Management:
- Empiric first line: Piptazo 4.5g IV q8H OR Ceftazidime 2g IV q8H
- Add Vancomycin 15mg/kg IV q12H if evidence of skin/soft tissue OR line infection OR extensive mucositis OR septic shock with respiratory distress OR MRSA
- Add Flagyl for concerns of intra-abdominal infection if using Ceftazidime alone
- Add respiratory fluoroquinolone or azithromycin if suspecting Legionella in severe respiratory infection
- Please see order sets as protocol may vary according to site

SPENECTOMIZED PATIENTS

Assessment:
- How did they lose their spleen?
  - History of prior surgical splenectomy
  - "Autosplenectomy" from sickle cell or collagen vascular disease
- Confirm suspicions with Howell-Jolly bodies on blood smear

Management:
- **Start antibiotics ASAP if concerned for infection!** Do not wait for cultures.
- Vancomycin 15mg/kg IV q8-12h + ceftriaxone 2g IV daily is a safe initial choice
- Use levofloxacin/moxifloxacin + vanco if pen allergic.
- Increase ceftriaxone to 2g IV bid for meningitis.
- Asplenic patients more prone to encapsulated organisms *(S. pneumo, N. meningitidis, H. influenza)*
- Review vaccination history
- See order set for Post-Splenectomy Vaccination if not up to date

FUNGEMIA / CANDIDEMIA

Assessment:

- Risk Factors:
  - Usually occurs in hospitalized and immunocompromised patients, intraabdominal surgery or perforation, chronic parenteral nutrition (TPN), central lines, neutropenia, candida infection from other sites, IVDU
- Most common organism will be *Candida*. Species **resistant to fluconazole** include: *Candida glabrata and Candida krusei*
- **Note Candida in sputum and urine (often from patient with catheter) often is colonizer, not infection*

Management:

- Empiric Antifungal: Echinocandin – formulary at HHS Anidulafungin 200mg loading then 100mg daily; formulary at St. Joseph's Caspofungin 70mg loading dose then 50mg daily (avoid caspofungin in patients with moderate-severe hepatic impairement), but also has Anidulafungin
- Fluconazole 800mg (12mg/kg) loading dose then 400mg (4mg/kg) daily can be considered in non critically ill patients who are unlikely to have a fluconazole-resistant *Candidia* species
- Treat for minimum 2 weeks from first negative blood culture and narrow once susceptibilities are back (usually step down echinocandin to fluconazole within 5-7 days)
- MUST remove lines
- Consult ID
- Consult Opthamology to rule out endopthalmitis

FEVER OF UNKNOWN ORIGIN (FUO)

Definition:

- Temperature greater than 38.3 that lasts for more than 3 weeks with no obvious source despite appropriate investigation
- Categories:
  - Classic: evaluation of atleast 3 outpatient visits or 3 days in hospital
    - E.g. Infection, malignancy, collagen vascular disease
  - Nosocomical: patient hospitalized greater than 24 hours with no fever on admission
    - E.g. C. Diff, drug-induced, pulmonary embolism, septic thrombophlebitis, sinusitis
  - Immune deficient/neutropenic: neutrophil count <= 500 per mm3
    - E.g. Opportunistic bacterial infections, aspergillosis, candidiasis, HSV
  - HIV associated: duration of >4 weeks outpatient, >3 days inpatient with confirmed HIV
    - E.g. CMV, MAC, PJP, drug-induced, Kaposi's sarcoma, lymphoma

Assessment:

- Differential Diagnosis of FUO that may be missed:
  - o Infection: Abscesses (abdominal, pelvic, dental), Endocarditis, Osteomyelitis, Sinusitis, TB (inc. extrapulmonary sites), CMV, EBV, HIV, Lyme, Prostatitis, Hepatitis
  - o Malignancies
  - o Autoimmune: Adult onset Still's disease, PMR, Temporal arteritis, RA, IBD, Reactive Arthritis, SLE, Vasculitis
  - o Others: Drug-induced, DVT/PE, Sarcoid
- Investigations:
  - o CBC, lytes, extended lytes, Cr, BUN, VBG, CRP/ESR, Blood Culture x2, Urine culture, CXR, CT, stool studies, virology
- Common causes of leukocytosis >20 to be considered:
  - o C.diff, intra-abdominal abscess, fungemia, malignancy

ANTIBIOTICS

"GP" = gram positive and "GN" = gram negative

Note that guidelines and clinical practice often differ, and evidence is not always available to back up antibiotic recommendations. Consider each patient individually and where the source of infection, severity of illness, risk factors for resistant organisms, previous microbiology, and local resistant patterns.

Antibiograms are available on HHS intranet for susceptibility patterns.

Principles:

- **Allergies**: Differentiate "true allergy" from "adverse drug effect" (eg. GI upset). Only 6-10% of first generation cephalosporins cross-react with a history of penicillin allergy and +ve skin test (lower with higher generations). But seek advice from Infectious Disease before using cephalosporins in a patient with history of life-threatening penicillin allergy. Consider allergy testing or exposure therapy.
- **Broad vs. Narrow Spectrum**: Pick the fewest drugs with the widest coverage. When culture results are back, or clinical situation changes, narrow the spectrum.
- **Drug Effects:**
  - o Renal Toxicity: Calculate initial creatinine clearance and follow closely. Adjust drug dose and/or frequency if necessary. Be wary of aminoglycosides!
  - o Venous Access: Poor access may necessitate simpler regimen or oral/IM route. Quinolones have excellent oral bioavailability, so give orally whenever possible. Very sick patients with third spacing/gut edema may have difficulty absorbing oral agents.
  - o Bacteriostatic vs. Bacteriocidal: Serious infections require antibiotics which have a bacteriocidal mechanism of activity (ex. Beta-lactams, quinolones, vancomycin, metronidazole)

ANTIBIOTIC DOSING

Class	Antibiotic	Dosing	Renal adjustment
Penicillins	Penicillin G	2-4 M units IV q4-6	CrCl 10-50: 75% dose
	Penicillin V	250-500 mg po tid/qid	Nil
	Cloxacillin	2g IV q4-6 hrs	Nil
Amino-penicillins	Ampicillin	1-2 g IV q4-6h	CrCl 10-50: q6-12h
	Amoxicillin	250-1000 mg po tid	CrCl 10-30: 250-500q12h
	Amoxicillin-Clavulanic Acid	875/125 mg po bid	CrCl 10-30: 250-500q12h
Anti-pseudomonal penicillins	Piperacillin Tazobactam*	4.5 g IV q8h if pseudomonas q6h	CrCl < 15: 2.25 q8h
Carbapenems	Imipenem*	500-1000 mg IV q6h	Adjust to CrCl
	Meropenem*	1g IV q8h (use 2g if CNS dosing)	CrCl 25-50: 1g q12h CrCl 10-25: 0.5g q12h CrCl <10: 0.5g q24h
	Ertapenem	1g IV q24h	CrCl <30: 500g q24h
	Imipenem*	500-1000 mg IV q6h	Adjust to CrCl
Cephalosporins			
First gen	Cefazolin	1-2g IV q8h	CrCl 11-34: 50% dose q12h
	Cephalexin	250-1000 mg po qid	CrCl 10-50: 500 mg q8-12h
Second gen	Cefuroxime	125-500 mg po bid OR 750-1500mg IV q8h	CrCl 10-30: dose q24h
Third gen	Ceftriaxone	1-2g IV q24h 2g IV q12h in CNS	Nil
	Ceftazidime*	1-2g IV q8-12h	CrCl 30-50: 1g q12h CrCl 15-30: 1g q12h
Aminoglycosides (caution with renal and ototoxicity)	Gentamicin*	5-7mg/kg IV q24h	Adjust to CrCl
	Tobramycin*	5-7mg/kg IV q24h	Adjust to CrCl
	Amikacin*	7.5 mg/kg q12h	Adjust to CrCl
Fluoro-quinolones	Ciprofloxacin*	500-750 mg po bid OR 400 mg IV	CrCl 30-50: 250-500 q12

		bid	CrCl 5-30: 250-500 q18
	Moxifloxacin	400 mg po/IV daily	Nil
	Levofloxacin*	500-750 mg po/IV daily	CrCl > 20: nil CrCl 10-19: 250 q48h
	Norfloxacin	400 mg po bid	CrCl < 30: 400 daily
Macrolides	Azithromycin	500mg load then 250 mg po/IV daily	Nil
	Clarithromycin	250-500 mg po q6-12h	CrCl < 30: decrease dose by 50%
	Erythromycin	250-500mg po q6-12h	CrCl <10: decrease dose by 50% ·
Tetracyclines	Doxycycline^Δ	100 mg po q12h	Nil
	Tetracycline	500 mg po qid	CrCl 50-80: q8-12h CrCl 10-50: q12-24
	Tigecycline^Δ	100 mg IV, then 50 mg q12h	Nil
Sulfa	Trimethoprim-Sulfamethoxazole^Δ	1-2 SS/DS tabs po bid 8-20mg/kg/day IV q6-12h	CrCl 15-30: 50% of dose CrCl <15: full dose q48h
Clindamycin	Clindaymycin^Δ	150-450 mg po qid or 300-600 mg IV q6-12h	Nil
Glycopeptides	Vancomycin^Δ	15 mg/kg IV q12h	Dose based on weight and CrCl
Oxazolidinones	Linezolid^Δ	600 mg po/IV q12h	Nil
Lipopeptides (Mandatory ID Consult)	Daptomycin^Δ	4-6 mg/kg q24h Use up to 8-12 mg/kg in MRSA	CrCl <30: dosing q48h
Antifungals	Fluconazole	12mg/kg loading dose, then 6mg/kg daily (800mg then 400mg)	CrCl <50: reduce dose by 50%
	Caspofungin	70mg IV x 1, then 50 mg IV daily	Nil
	Anidulafungin	200mg IV x 1, then 100 mg IV daily	Nil
Antivirals	Acyclovir	10mg/kg IV q8H	CrCl 25-50: q12h CrCl 10-24: q24h CrCl <10: 50% dose q24h

*pseudomonal coverage
^ΔMRSA coverage

RESPIRATORY FAILURE - ACUTE

Definitions: Type I respiratory failure = failure to oxygenate (low pO_2)
Type II respiratory failure = failure to ventilate (high pCO_2)
In practice, usually a mixed picture and seldom in isolation

Common Clinical Scenarios:

- COPD exacerbation
- Cardiogenic pulmonary edema (pump failure) vs. non-cardiogenic pulmonary edema (ARDS, neurogenic, TRALI)
- Pneumonia vs. aspiration pneumonitis
- Asthma exacerbation
- Pulmonary embolism
- Drugs (esp. opioids, benzodiazepines)
- Pneumothorax
- Generalized allergic reaction, angioedema

Diagnostic Approach:

1. Assess ABC's: vitals (specifically SpO2 and RR), distress level (posture, acc. muscle use and muscle fatigue), LOC, cyanosis.
2. Ascertain baseline pulmonary function: best FEV1, past admissions to ICUs requiring non-invasive ventilation/mechanical ventilation (including airway assessments), ABG's (previous PaCO2s), functional ability.
3. Look for precipitating factors: infection, heart failure/COPD precipitants, recent thoracic procedures (pneumothorax), narcotics
4. Other significant medical conditions affecting ventilation/perfusion, cardiovascular, neuro (eg. Guillan-Barre or Myasthenia Gravis), malnutrition, ascites
5. Resp exam: determine the pattern of breathing (rapid, shallow, paradoxical); inspection, palpation, percussion, auscultation; check for chemosis (conjunctival edema), asterixis; inspect O2 delivery system.
6. Investigations: CBC, lytes, Cr, ABG, ECG, CXR, sputum C+S, blood culture if febrile or chills/rigors.

Oxygen Therapy for Hypoxia:

1. Empirically give O2 by mask to achieve sats over 90%, AVOID nasal cannula in acute settings. Titrate oxygen down to the minimum needed for sats over 90% as soon as possible.
2. If O2 suppl ineffective and no immediate contraindications, can trial BiPAP or Optiflow/Airvo (High Flow Nasal Cannula). (for severe viral/community acquired pneumonia in isolation, use HFNC; all other conditions, trial BiPAP)

Ventilatory Therapy for Hypercarbia: AVOID respiratory suppressants (opioids/benzos/antipsychotics) including over-oxygenation (goal sats 88-92%); use either non-invasive or invasive ventilatory supports (see below)

Treat Underlying Causes:

1. COPD or Asthma:
   - *Bronchodilators:* Atrovent and ventolin 1neb back-to-back X 1h if severe; can switch to Atrovent and Ventolin 2-4 puffs q4h + ventolin 1-2 puffs q1hr prn as sx improves
   - *Steroids:* dose and route of administration based on disease severity. In most severe cases, IV Solumedrol 60-125 mg q 6hr; in mild cases, po steroid 0.6 – 1 mg/kg OD.
2. Pneumonia: empirical abx Rx, see "Sepsis" section.
3. Pulmonary edema: nitro patch 0.4mg or furosemide IV; 40 mg if normal Cr and 80-160 mg if renal fxn impaired; repeat clinical evaluation +/- ABG and CXR in 4-6hrs.
4. Maintain high suspicion for PE: start heparin IV or LMWH if no contraindications and no other obvious cause of respiratory failure, consider thrombolytics if hemodynamic instability – discuss with your SMR stat for use of TPA
5. GERD, aspiration pneumonia: Gaviscon, domperidone, ranitidine or PPIs.
6. Pneumothorax: chest tube required, call your SMR, place patient on O_2
7. Narcotic Overdose: narcan 0.4 mg IV "stat"; be aware short half-life of narcan and need for repeat doses or infusion

Clinical Warning Signs (indicate need for intubation):

- Main questions
  1) Failure protect airway?
  2) Failure to oxygenate?
  3) Failure to ventilate?
- Decreased LOC
- Silent chest
- Respiratory muscle fatigue; abdominal paradox
- Normal or increased pCO2 with tachypnea suggesting respiratory fatigue
- Transition from tachypnea to bradypnea
- Pulsus paradoxus
- Prolonged FiO2 requirement > 0.40; inappropriate pO2 response to O2 Rx (i.e. pO2 rise < 2 mmHg with 1% incr in FiO2)
- Signs of myocardial ischemia: chest pain, pulmonary edema, arrhythmia, ischemic ECG changes, hypotension
- Uncontrolled secretions/suspected mucous plugging

Indications:

- Airway protection and control
  - GCS<8
  - Relief of airway obstruction: eg. smoke inhalation/angioedema
  - Airway protection: eg. vomiting and decreased LOC = risk of aspiration
  - Facilitate bronchial hygiene: eg. weak cough = inability to clear secretion
- Failure ventilate (and failed NIV or not appropriate for NIV)
  - Mechanical ventilation: eg. resp failure refractory to above Rx
  - Clinical resp distress: abdominal paradox, large pulsus paradox, silent chest, RR > 35, decreased LOC, cannot finish a word
  - ABG: $PaCO_2$ > 50 mmHg and pH < 7.30
- Failure oxygenate – PaO_2<60 or FiO_2>0.6
  - Diffusion deficits
  - V/Q mismatch or shunt
  - Dead space ventilation
- Route for drug administration: eg. in code situation when IV not available
- Minimize O_2 consumption/agitation: post-MI, shock, raised ICP

Equipment:

- 'SOLE': suction, oxygen, laryngoscope, ETT (male: size 8; female 7-8; minimum size 8 is required for bronchoscopy)
- Others: cardiac monitor, bag valve mask, oral airway, stylet, gloves, tape, water soluble lubricant, 10 cc syringe, stethoscope, airway rescue supports (LMA, Glidescope, bougie, tracheostomy tray/Melker kit)

Medications:

- Lidocaine spray: 10 mg/spray, max 20 sprays
- Sedatives (in order of most to least common): Propofol 0.5-1mg/kg IV initially then 20 mg q10 sec (onset 40 sec, duration 5-8 min); Ketamine 0.75-1mg/kg IV initially then 30 mg alloquots q1 min (onset 30 sec-1 min, duration 10-15 min; Ketamine can often cause bronchospasm and requires NM blockade at the bedside) midazolam 0.03-0.08 mg/kg IV initially, then titrate q2-5 min (onset 2-6min, duration 0.5-2hr)
  - Sedatives will drop BP – have phenylephrine at bedside PRN (100 - 200 mcg boluses)
- Analgesics: fentanyl 0.5-1mcg/kg/dose (onset 4min, duration 30-45min)
- Neuromuscular blockers: rocuronium 0.5-1mg/kg/dose; succinylcholine 1-1.5mg/kg/dose; AVOID NM blockers unless airway expert present.
- Vasopressors: phenylephrine – mix 10mg in 100ml NS, draw 10ml mixture (1 mg) and give 1-2ml (0.1 mg or 100 mcg) q5 min for post-intubation hypotension.

Intubation procedure:

Most important: anticipate difficult airway (neck trauma, c-spine injury, < 3 finger breath mouth opening, large tongue, thick neck, unable to visualize uvula or tonsillar pillars), call anaesthesia for back up

1. Pre-oxygenation with mask airway +/- oral airway +/- NIV +/- nasal cannula
2. Elevate head 10 cm with pads/pillows under occiput
3. Head extension at atlanto-occipital joint to allow for **sniffing position (r/o c-spine injury first)**
4. Adjust bed to your own sternum
5. Hold laryngoscope with left hand
6. Enter pt's month via right side and sweep tongue to left
7. Advance blade to base of tongue (i.e. vallecula – space between base of tongue and epiglottis)
8. Lift laryngoscope to expose vocal cords
9. Have RT apply cricoid pressure or BURP if needed (Sellick manoeuvre)
10. Pass ETT on inspiration (cords open) and remove stylet just as the tip of the tube passes the vocal cords. Insert ETT to a distance from the lip of 20 cm for female and 22 cm for male.
11. Inflate cuff with 10cc of air and tape ETT securely
12. Verify ETT position (below)

Verification of ETT Position: (ideal is using multiple mechanisms)
- Direct visualization of ETT tube passing through vocal cords
- Auscultation for equal and bilateral breath sounds
- Condensation in ETT tube
- Capnography
- Clinical exam: resolution of hypoxia and improvement in vital signs
- Oximetry and ABG
- CXR: tip of ETT should be 1-2 cm above the carina.

Post Intubation Infusion Drugs:
- Sedatives: **propofol** 0.3-1mg/kg IV bolus then 0.5-3mg/kg/hr IV infusion, **midazolam** 0.01-0.1mg/kg IV bolus then 0.5-7 mg/hr
- Analgesic: **fentanyl** 1-2 mcg/kg IV bolus then infuse at 1-3 mcg/kg/hr
- NM blockers: **cisatracurium** 0.1 mg/kg IV bolus then 3 mcg/kg/min, **rocuronium** 0.5-1mg/kg IV bolus then 10-12mcg/kg/min infusion (consult with ICU Senior/Fellow/Staff before implementing NM blockade)
- Stress Ulcer Prophylaxis: **pantoloc** 40 mg IV/PO q24 hr or ranitidine 50 mg IV q12hr
- **Fragmin** 5000 units s/c OD for DVT prophylaxis; TEDs.
- VAP Prophylaxis: **chlorhexidene** mouth wash QID

Complications of Intubation:
- Trauma during intubation: dental damage, upper A/W structure damage, aspiration, vagal stimulation and hypotension/bradycardia, arrhythmias, bronchospasm, pneumothorax, mediastinal/subcutaneous emphysema, laryngeal edema
- Complications with ETT in place: esophageal intubation, bronchus intubation, tube occlusion (thick secretion, kink, biting), cuff leak, laryngeal or tracheal granuloma, dislodgement, VAP

INVASIVE MECHANICAL VENTILATION

Common conditions leading to mechanical ventilation:

- Acute pulmonary parenchymal disease: ARDS, pneumonia, aspiration, CHF
- Acute airway disease: asthma, COPD exacerbation, inhalation injury
- Primary ventilation failure: GBS, MG, drug OD, chest wall diseases or injury
- Systemic disease: DKA, undifferentiated shock, sepsis
- Acute agitation, trauma, status epilepticus
- Peri-operative

Modes of IPPV (invasive positive-pressure ventilation):

1. **Volume-controlled (seldom used)**: delivers a preset volume of gas with each machine breath
    a. **Control Mode**: delivers a set rate and Vt regardless of pt effects (rarely used)
    b. **Assist control (AC)**: pt triggers machine (with backup rate), which delivers a set Vt; pt can breath faster (with full Vt support) and above the Vt; AC is best used in paralyzed and apneic pts - spontaneously breathing pt with a high ventilatory drive often continues to have a high work of breathing.
    c. **Synchronized intermittent mandatory ventilation (SIMV)**: machine delivers intermittent mandatory preset tidal volume at a preset interval; this mandatory breath is synchronized with the patient's spontaneous breathing. When pts breathe above the preset rate, they do so without ventilator support.
2. **Pressure-controlled**: delivers a preset pressure with each machine breath
    a. **Pressure control (PC)**: delivers a set rate of breath by giving pts a preset positive pressure (Pi) for a preset duration (Ti); pts may breath faster with full support and generates more inspiratory negative pressure; goal is to limit peak airway pressure to < 35-40 cmH2O; good for barotrauma, or use in inverse I/E ratio ventilation.
    b. **Pressure support (PS)**: **pt triggered breaths** with positive inspiratory pressure support, without Ti or backup rate. NOTE: CPAP + PS in IPPV is equivalent to BiPAP in NIPPV
3. **Positive end expiratory pressure (PEEP):** recruits previously collapsed alveoli and holds then open during expiration; counteracts airway resistance of tubing; impedes venous return and decreases afterload. May improve LV dysfunction or may cause hypotension if very preload dependent.
4. **High frequency ventilation (HF):** jet ventilation and oscillator (both RARELY used), latter draws air out of lungs; used for severe hypoxemia as a rescue therapy; often (but not always) requires chest tube and paralysis
    Note: If at a site with RTs onsite, consult with thembefore making ventilator setting changes

Initial Ventilator Settings:

- Fi02: 1.0, aim to decr below 0.6
- Vent rate: 10-20, dependent requirement of hyperventilation
- Vt: 6-8 ml/kg (if in pressure control mode, need to monitor resultant volumes)
- I:E ratio (inspiration time:expiration time): 1:2
- PEEP: 5-15 cmH2O (depending on clinical condition and body habitus)
- PIP: < 35 cmH2O
- Mode: PCV

Adjusting PaO2:

1. To decrease: decrease FiO2 by 0.1 to 0.2 q30min; remember "Rule of Seven" – for every 1% decrease in FiO2, PaO2 decreases by 7 mmHg
2. To increase:
   a. Increase FiO2 by 0.1 to 0.2 q30min
   b. Increase PEEP by 2-4 cmH2O q30min, but monitor cardiac output.
   c. Increase ventilation if PaCO2 high

Adjusting PaCO2:

1. To decrease:
   a. Increase minute-ventilation by increasing RR and/or Vt
   b. Check system air leak
2. To increase:
   a. Decrease minute-ventilation by decreasing RR and/or Vt. Aim for Vt in the 5-6ml/kg of ideal body weight range.
   b. Switch from AC to SIMV if pt is hyperventilating
   c. Increase exhalation tubing to increase dead space

Weaning Protocols:

- **PCV → PS** → reduce pressure by 5cmH2O while monitoring ABG and RR. If RR>25, then pressure should be increased. Good for reconditioning of resp muscles. Once pressure reaches 5-10, then ready to extubate.
- **AC → CPAP or T-piece** if short intubation in otherwise healthy pt.
- **SIMV** → decr backup rate by 2-4/min while monitoring ABG; if RR >25, then resp muscle fatigue, and backup rate should be increased. Good for reconditioning of resp muscles. Once backup <= 4, then ready to extubate
- **NIV** can be used to help facilitate extubation

Extubation Indications:

- Systemic Readiness
  - Adequate oxygenation: PaO2 >60 mmHg on FiO2<0.4; PaO2/FiO2 >150–300
  - Stable vitals: good BP with no (or minimal) pressors, afebrile
  - Stable metabolic status: normal acid-base, acceptable renal function and electrolytes, including extended lytes.
  - Adequate Hb: >80–100 g/L
  - Adequate mentation: rousable, GCS >13, no sedative infusions
  - Others: adequate pain control, nutrition, rest and sleep
- Ventilatory Readiness
  - Properly weaned: T-piece, CPAP, SIMV backup ≤ 4/min, PS 5-10, or PS 5-7cmH2O + PEEP 5cmH2O; RR ≤ 25.
  - A/W: cuff leak (exp easier to achieve than insp)
  - Respiratory muscle: MIP (max inspir pressure) < -20 to -30 cmH2O
  - RSBI (rapid shallow breathing index = resp rate to tidal volume) < 85-100
  - VC = 10-15 ml/kg
  - Minute Ventilation <10
  - Strong cough during suctioning

Extubation failure: retry in 24h; 10% re-intubation rate is acceptable; many deserve a trial of extubation despite suboptimal parameters

NON-INVASIVE MECHANICAL VENTILATION

Types:

- **HFNC (high flow nasal cannula) AKA Airvo or Optiflow**
  - o provides 4-6cm H2O of PEEP (in theory, stents open smaller airways promoting oxygenation)
  - o best studied in patients with severe pneumonia to prevent intubation
- **CPAP (continuous positive airway pressure)**: same as PEEP (~10cm H2O) throughout resp cycle
  - o Elevates baseline pressures
  - o Improves oxygenation by keeping airways open and recruited
- **BiPAP (bilevel positive airway pressure)**: inspiratory and expiratory pressures
  - o Positive pressure in inspiration decreases WOB by unloading diaphragm
  - o IPAP and EPAP synchronized with resp cycle
    - ▪ IPAP – pressure support that reduces WOB and thus O2 consumption
      - Improves ventilation – decreases resp muscle effort, increases TV and alveolar ventilation
    - ▪ EPAP – prevents collapse and maintains positive pressure in airways
      - Improves oxygenation – increases SA of alveoli = improved gas exchange

Indications:

- Hypoxemic respiratory failure: HFNC (pneumonia) or BiPAP (CHF/AECOPD)
- Hypercarbic respiratory failure (eg. COPD): BiPAP
- Mixed respiratory failure: BiPAP
- Pts who refuse intubation
- Post extubation in difficult to wean pts

Practical points:

1. Place patient in stepdown, CCU or ICU; may initiate NIV on ward while patient is being arranged for transfer in some cases
2. Patient must be relatively alert and cannot have excessive secretions or continuous vomiting (risk of aspiration).
3. Head of bed > 30 degrees
4. Selection of masks: facial masks preferred over nasal mask, as most pts in acute respiratory failure are mouth breathers
5. Start CPAP setting at 10 cmH_2O, may need more for obese patients; may titrate up to 20 cmH_2O
6. Start BiPAP setting at inspiratory pressure 15 cmH_2O and expiratory pressure 8-10 cmH_2O; may titrate up to 20 cmH_2O inspiratory pressure
7. Provide a backup rate of 8 per min.
8. Provide supplemental O_2 and humidity as needed
9. Recheck ABG in 1-2 hours
   - Predictors of success:
     - o Decrease PaCO2 >8mmHg, increase pH >0.06, correction of resp acidosis
   - Of failure:
     - o Illness severity (pH<7.25, PCO2>80), decreased LOC, unable tolerate mask/clear secretions
   - Expect to see results at 2 hours

10. Use sedation VERY carefully, only in extremely agitated pts (eg. lorazepam 0.5 mg PO)
Note: As with mechanical ventilated patients if RTs are onsite, consult with them before making ventilator setting changes

Relative Contraindications:

- Unable to protect airway; high risk of aspiration
- Upper airway obstruction or facial trauma
- Unable to clear secretions
- Hypotension, hemodynamic instability
- Unable to cooperate or tolerate mask ventilation
- Uncontrolled arrhythmia
- GI bleed
- Evidence of coronary ischemia

Absolute Contraindications:

- GCS <8
- Cardiac, respiratory arrest

Complications:

- Gastric distension, facial pressure sores, dry MM with thick secretions, aspiration, barotrauma
- Hypotension – preload reduced as intrathoracic pressure becomes more positive

SHOCK

- Reduced systemic perfusion → diminished O2 to tissues → organ damage
- Definition requires hypotension + clinical/biochemical signs of tissue hypoperfusion
  - **Hypotension** – varies but generally SBP<90, MAP<65
  - **Signs of tissue hypoperfusion** – cyanosis, clammy skin, decreased UO<0.5mL/kg/mn, altered LOC
  - *High lactate >4, metabolic acidosis = altered cellular metabolism*
- **ENSURE CAB'S Call CODE if necessary.** Stat O_2, cardiac monitor, Trendelenburg position, IV 14-16 gauge X2, will need a central line if require inotropes/vasopressors.
- **DETERMINE CODE STATUS**
- Fluid challenge **crystalloid (RL PREFERRED) 1 litre bolus** while assessing patient (if signs of pulmonary edema continue with fluid but consider need for mechanical ventilation)
- Clinical history and latest blood work.
- Assess vital signs. Check pulsus paradoxus if suspect tamponade.
- Assess JVP (elevated in PE, cardiogenic, tamponade and usually low in others)
- CVS, Resp (assess for tension pneumothorax and for pulmonary edema), Abdo, peripheral exam
- Assess rhythm → ECG
- Stat CXR. Intubate if necessary.
- End organ → cyanosis, mental status, urine output, feels hot or cold.

Differential Diagnosis:
- **Hypovolemia**: hemorrhage, trauma, 3rd spacing (esp. post-op), pancreatitis, G.I. losses, burns, diuretics, post-ATN diuresis.
- **Cardiogenic**: MI, arrhythmias, severe cardiomyopathy, acute septal or valvular rupture, pericardial tamponade, P.E., tension pneumothorax.
- **Distributive**: Sepsis and Anaphylaxis
  - **Neurogenic**: head trauma, CNS/spinal lesions (sympathetic tones), Vasodilator excess.
  - **Acute adrenal insufficiency**: prior steroids with acute stress, endocrine Hx.

Management:
- Goals for septic shock
  - **MAP >65**
  - **UO >0.5mL/kg/mn**
  - **Central venous O2 sat >70%** (seldom assessed now)
- Bloodwork: Routine plus: cross + type 4 units, INR, PTT, CK/Trops, Ca, albumin, amylase, lactate, ABGs
- ?Infective focus: Blood cultures, sputum, urine, lines. Initiate **broad spectrum IV antibiotics.**
- Crystalloid fluids wide open, **target MAP 60-65, sBP > 100 mmHg**. ***Consider albumin after 3L crystalloid.*** PRBC stat for known hemorrhage. Consider FFP + Plts after 6U PRBC.
- Maintain urine output → Insert Foley catheter.
- Acidosis: Consider NaHCO3 IV for pH <7.0

- Adrenal insufficiency? Look for history of steroid usage. If in doubt, treat anyway, eg. **hydrocortisone 50mg IV q6h or 100mg IV q8h**. Can consider drawing random cortisol and ACTH first
- ICU? Transfer for prolonged episode, unstable. Consider calling **critical care response team** if available at your hospital. Arterial line, central line access for inotropes

Cause	CO	SVR	PCWP	EDV
Hypovolemic	-	+	-	-
Cardiogenic	-	+	+	+
Obstructive – Afterload	-	++	+	+
Obstructive – Preload	-	+	+	-
Distributive	+	-	-	-

Hemodynamic Indices in Shock

Cause	CVP	PCWP	CO	HR	SVR
Septic	+/–	–	++	+	–
Hypovolemic	–	–	+/=	+	++
Cardiogenic	+	+	–	+/=	+

Cardiogenic/Obstructive Shock

Cause	Rt Atrium	PCWP	CO	2D Echo
LV dysfx	+	++	– –	Poor LV
RV dysfx	++	–	– –	Dilated RV
Tamponade	++	++	– –	Tamponade
PM rupture	+	++	– –	Severe MR
VSD	+	++	+	VSD
PE	++	N	–	PADP>PCWP; dilated RV

INOTROPES + VASOPRESSORS

Inotropes: Increase cardiac contractility
Vasopressors: Induce vasoconstriction

- No guidelines for initiation of these drugs – start when MAP<60 or SBP decrease >30mmHg from baseline AND when unresponsive to initial crystalloid resuscitation
- Goals: To perfuse end organs (urine output, mentation, lactate), MAP 60-65/SBP 90-100 mmHg
- Useful when unable to maintain pressure despite adequate fluid resuscitation (require adequate volume resuscitation/intravascular volume to be effective)
- Monitor with arterial line

Adrenergic Receptor Activity

Receptor	Effects
Alpha-1 – vascular walls	Vasoconstriction
Beta-1 – heart	Inotropy, Chronotropy
Beta-2 – vascular walls	Vasodilation
Dopamine – renal, cerebral, splanchnic, coronary vasc	Vasoconstriction (Norepi release), Renal/splanchnic/coronary vasodilatation

Specific Agents

Drug	Alpha-1	Beta-1	Beta-2	Comments
Phenylephrine	+++	0	0	Vasoconstriction with reflex bradycardia Consider for tachyarrhythmias
Norepinephrine	+++	++	0	Vasoconstriction, modest incr CO. May induce tachycardia, reflex brady
Epinephrine	++	+++	++	Vasoconstriction, incr CO
Dopamine	0/+/++	+/++	0	No role for renal dose
Dobutamine	0/+	+++	++	Incr CO, decr SVR Risk of hypotension
Isoproterenol	0	+++	+++	Chronotropy, vasodilatation, decr MAP
Vasopression	n/a	n/a	n/a	ADH analogue
Amrinone, Milrinone	n/a	n/a	n/a	Inotropic, vasodilatory Milrinone decreases pulm HTN

Choice of Agents

		1ST LINE Pressor	2ND LINE Pressor
Septic shock		Norepinephrine	Vasopressin (Norepi sparing) Epinephrine
Heart failure		Norepinephrine	Dopamine/Milrinone
Cardiogenic shock		Norepinephrine	Dopamine
Anaphylactic shock		Epinephrine	Vasopressin
Neurogenic		Norepinephrine	Dopamine
OR Hypotension	Post-anesthesia	Phenylephrine	
	Post CABG	Epinephrine	Norepinephrine

*** If evidence of myocardial dysfunction (pump failure) you may need to consider an inotrope as well (dobutamine or milrinone)*

***Most pressors have some inotropic effect as well (epi > dopamine > norepinephrine)*

Complications:
- **Hypoperfusion** → to extremities, kidneys, mesenteric organs (gastritis, shock liver, intestinal ischemia)
- **Lactic acidosis** → due to metabolic effects (epi) or splanchnic hypoperfusion
  - Secondary to poor peripheral perfusion with vasoconstrictors
- **Dysrhythmias** → Beta-1 chronotropy (sinus tachy, AF, AVNRT, ventricular tachyarrhythmias)
- **Myocardial Ischemia** → inotropy increases myocardial demand
- **Local effects** → if peripheral line used may cause skin necrosis if IV goes intersitial
- **Hyperglycemia** → inhibit insulin secretion (esp with Epi)
- **Tachyphylaxis** → constantly reassess dosing and clinical condition
- **Inotropes** (milrinone, dobutamine) cause vasodilation and can drop BP before the inotropic effect begins – adding a pressor (usually norepinephrine) may minimize hypotension

Surviving Sepsis Guidelines:
- **Norepinephrine** is first choice pressor
  - Add in an additional agent if needed to maintain MAP >65
- **Vasopressin** (0.03 units/kg/hr) can be added to NE to either raise MAP or decrease NE dose
  - Low dose vasopressin not recommended as single initial pressor for sepsis-induced hypotension
  - Doses higher than 0.03-0.04 units/kg/hr should be reserved for salvage therapy
- Can use dopamine as alternative to vasopressin in select patients (low risk tachyarrhythmia and absolute or relative bradycardia)
- Phenylephrine **not** recommended except if
  - NE associated with serious arrhythmias
  - Cardiac output is known to be high and BP persistently low
  - Salvage therapy
- Inotropic support
  - Trial **Dobutamine**, start at 5mcg/kg/min up to 20mcg/kg/mn if

- Myocardial dysfunction – high cardiac filling pressures and low CO
- Ongoing signs hypoperfusion despite adequate intravascular volume and MAP

Dosage Range and Indications for Selected Vasoactive Drugs:

Drug	Usual Range	Maximum*	Common uses
Phenylephrine	20-200 mcg/min	360 mcg/min	Septic shock, neurogenic shock, post-anaesthesia-induced hypotension
Norepinephrine	0.5-20 mcg/min	30 mcg/min	Septic shock, cardiogenic shock
Epinephrine	2-10 mcg/min Anaphylaxis: 0.3-1 mg	20 mcg/min	Anaphylaxis, ACLS, septic shock, cardiogenic shock
Dopamine	Dopaminergic: 1-2 mcg/kg/min Beta: 5-10 mcg/kg/min Alpha: 10-20 mcg/kg/min	50 mcg/kg/min	Septic shock, cardiogenic shock
Dobutamine	2.5-20 mcg/kg/min	40 mcg/kg/min	Cardiogenic shock
Isoproterenol	1-10 mcg/min	20 mcg/min	Cardiogenic shock with bradycardia
Milrinone	Load: 50 mcg/kg/10 min Maintenance: 0.375-0.75 mcg/kg/min	0.75 mcg/kg/min OR 1.13 mg/kg/day	Cardiogenic shock

**Maximum doses are not recommended routinely. They represent the maximum doses cited in the literature estimated where necessary for 70 kg patient*

CENTRAL VENOUS CATHETER

Indications
- Lack of peripheral sites
- For hypertonic solutions/meds
- Monitor central venous pressure
- Access for temporary pacemaker, Swan-Ganz Catheter, or venovenous hemodialysis

Contraindications
- Infection or burn at site
- Inability to find landmarks – try U/S guidance first
- Thrombosis of vein site
- Coagulopathy or Platelets <50 000

Complications:
- Carotid puncture or thoracic duct laceration if left IJ or subclavian line
- Pneumothorax or Hemothorax
- Thrombo, air, wire, or catheter embolism
- Infection, particularly after 3-4 days in a septic patient. The risk is femoral > IJ > subclavian. If site not infected looking & patient has no evidence of infection, then don't need to change q 5 days.

Procedures

***all central lines (except potentially during code blue) should be inserted using ultrasound guidance in order to minimize complications**

***Subclavian Vein*:** Insert needle 1 fingerwidth below the clavicle, 2-3 cm medial from the mid-point of the clavicle (where the clavicle starts angling posterior). Keep the needle as parallel to the skin as possible to avoid a pneumothorax, aim towards the sternal notch, hit the clavicle, walk the needle underneath, & advance needle along the posterior border of the clavicle until the vein is entered. Turn the needle, so that the bevel is toward the heart.

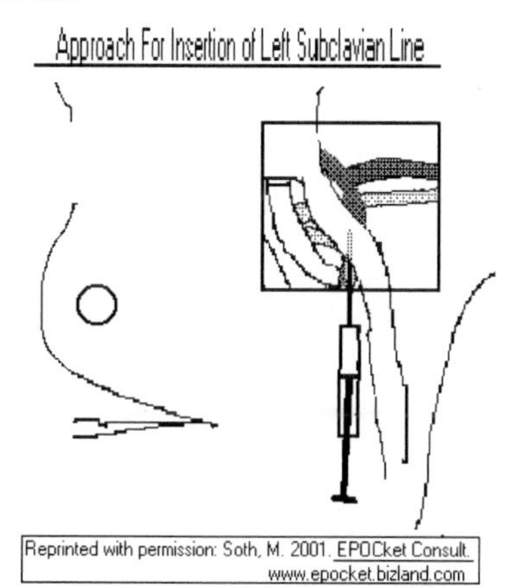

IJ/EJ – U/S guided

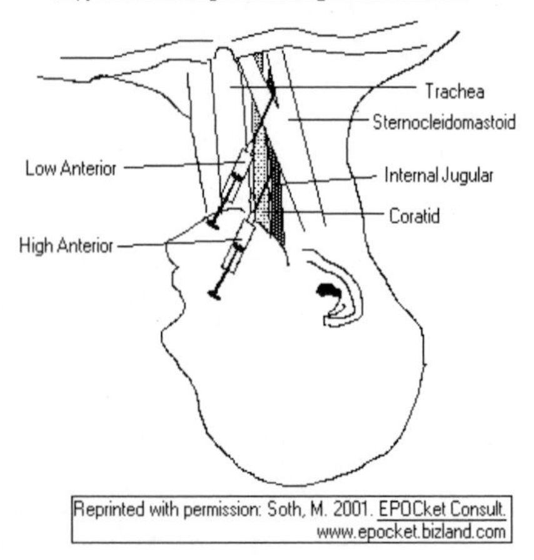

Femoral Vein: An easy approach medial to the femoral artery & inferior to the inguinal ligament. However, increased risk of thromboembolism or infection. Useful if all that is needed is central vascular access or hemodialysis, particularly in patients with coagulopathy or severely compromised respiratory status (won't tolerate a pneumothorax).

NEPHROLOGY

ACUTE KIDNEY INJURY (ACUTE RENAL FAILURE)

AKI Classification Systems used by KDIGO include RIFLE and AKIN.

Definition:

1) ↑Cr by 26.5 umol/L within 48h
2) ↑Cr 50% above baseline within 7 days
3) Oliguria (<0.5mg/kg/hr) for 6 hours

Tips to Creatinine Interpretation:

1. Compare current creatinine to previous values. **Check ClinicalConnect** for community bloodwork, which more likely reflects baseline renal function.
2. Determine the acuity of creatinine elevation to determine the timing to renal insult.
3. Consider the slope of creatinine rise and urine output. Creatinine improvement lags behind true renal recovery. Urine output may be more reflective acutely.
4. If the creatinine is unstable, it cannot be used to estimate GFR (Don't use for medication dosing!).
5. If creatinine is stable the CKD-EPI equation is preferred for estimating renal function
6. Recognize that creatinine is a breakdown product of muscle, so you may need to interpret it in the context of the patient's muscle mass (eg large muscular person vs cachetic patient will have different serum creatinines for the same renal function).

Differential Diagnosis:

	Etiologies	Clinical Findings/Workup
Prerenal	**Volume depletion**: blood loss, GI loss (diarrhea/vomiting), renal loss (diuretics, osmotic), skin losses (burns, sweating, insensible losses) **Ineffective circulation**: CHF, hypoalbuminemia (protein-losing, nephrotic, cirrhosis, malnutrition), 3rd spacing (pancreatitis, post-op), sepsis **Renal hemodynamics**: Drugs (NSAIDs, ACEI/ARB, cyclosporine, tacrolimus, amphoB, dye), ↑Ca, RAS, renal vein thrombosis, vasculitis	Volume status (hypovolemia, hypervolemia) Bland sediment with hyaline casts UNa < 20, FeNa < 1%, FeUrea < 35% UOsm > 500 Urea:Cr Ratio >10:100

Renal	**Glomerular Disease: Glomerulonephritis** ANCA (GPA, MPA, eGPA) Anti-GBM Immune Complex (PIGN, MPGN, IgAN/HSP, endocarditis, cryo, SLE)	Hematuria, HTN, edema Etiology specific: hemoptysis/sinusitis (MPA/GPA), asthma/mononeuritis multiplex(eGPA), recent infxn (PIGN), lupus features, new murmur (I.E.), abdo pain/purpura (HSP) Urine: dysmorphic RBCs, RBC Casts ACR or 24-hr urine collection (Subnephrotic proteinuria) ANCAs, anti-GBM, Anti-ASO, C3/C4, ANA, dsDNA, BCx, SPEP, HepC
	Acute Tubular Necrosis (ATN): Prerenal AKI: Ischemia from prolonged prerenal injury Endogenous toxins: rhabdo (myoglobin), hemoglobin, myeloma chains, uric acid crystals Exogenous toxins: contrast dye, meds (AmphoB, aminoglycosides, cisplastin), crystals (methotrexate, acyclovir)	CK, SPEP/Ca/HB/skeletal survey (myeloma), uric acid level Urine microscopy: muddy brown casts (75%), crystals FeNa > 2%, Urea:Cr < 10:100, UNa > 20, UOsm < 350
	Acute Interstitial Nephritis (AIN): Allergic (B-lactams, sulfa, NSAIDs, PPIs), infectious (pyelonephritis), infiltrative (sarcoid, leukemia, lymphoma), autoimmune (SLE, Sjorgen)	Urine microscopy: WBCs, WBC casts, RBCs, eosinophils (questionable utility) Biopsy if diagnosis uncertain
	Vascular: Polyarteritis nodosa, cholesterol emboli (post-cath), TMAs (DIC, TTP/HUS, HTN, scleroderma, APS)	Urine microscopy: eosinophils; INR, hemolysis w/u, APLA
Post Renal	**Lower GU obstruction (bladder neck, urethra)**: BPH, malignancy, **Lower GU neurogenic dysfunction**: DM, Parkinson's, anticholinergic meds, spinal cord injury	Bland Urine; Variable UNa/FeNa Bladder-scan, foley trial, renal ultrasound
	Bilateral ureteral obstruction: retroperitoneal fibrosis, lymphadenopathy, bilateral nephrolithiasis, malignancy	

Diagnostic approach:

1. History, General: oliguria, hematuria (multiple causes), edema/anasarca (glomerular disease)
2. History, specific pathologies (examples): Lower urinary tract symptoms (BPH, UTI), blood loss, dehydration, vomiting, diarrhea, heart failure, liver disease, hemoptysis (renal-pulmonary syndromes), social (HIV, Hep), B symptoms

(malignancy, indolent infection), autoimmune features, prolonged immobility or excessive exercise (rhabdomyolysis), contrast dye, sepsis (distribution)

3. History, drugs: Antimicrobials (B-lactam, sulfa, aminoglycosides, amphoB, acyclovir), immunosupressants (cyclosporine, tacrolimus, methotrexate), NSAIDs, PPIs, pamidronate, ACEI/ARBs, diuretics
4. Volume Status Assessment: Rule out hypovolemia, poor volume circulation (3rd spacing, cirrhosis, nephrotic syndrome, heart failure)
5. Other physical exam: heart failure (cardiac/resp), autoimmune findings (skin rashes, serositis, conjunctivitis, neurologic), vasculitis (purpura, mononeuritis)
6. Possible prelim. investigations: Urinalysis **(mandatory)**, urine culture (if infection suspected), urine lytes, , ECG (↑K), bladder-scan, renal ultrasound

Management:

1. Identify urgent dialysis indications: **A**cidosis (severe), **E**lectrolytes – hyperkalemia (refractory), **I**ngestions (Li, Methanol, Ethylene Glycol, salicylates, VPA), **O**verload of volume (refractory to diuresis), **U**remia (encephalopathy, pericarditis, uremic bleeding, N/V/pruritus)
2. Consider other indications for nephrology consult: severe AKI (Cr rise, anuria), RPGN
3. Foley catheter – 1) For accurate ins/outs (unless patient can urinate in container consistently); 2) Rule out urethral/bladder obstruction
4. Optimize fluid status – aim for euvolemia; Lasix +/- metolazone vs. IV crystalloid; bicarb infusion if significant acidosis
5. Treat electrolytes – hyperkalemia is common, sometimes hypokalemia if GI losses. There is no need to treat acute hyperphosphatemia.
6. Consider whether contrast-dye or gadolinium-enhanced imaging is necessary; possible increased risk of AKI and nephrogenic systemic fibrosis respectively
7. Stop all nephrotoxic medications – commonly diuretics (if not hypervolemic), NSAIDs, ACEi/ARBs, certain antimicrobials
8. Renally dose home medications and new medications in hospital, especially antibiotics. Doses may need to be changed as renal function fluctuates.

Contrast-induced AKI (CIAKI):

Definition: Cr rise of 25% or 26.5 umol/L over 48 hours. Classically peaks in 3-4 days and resolves after approximately 1 week.

Management:

1. Pre and post-hydration with IV Fluids: Consider IV crystalloid at ~50-100cc/h for 6-12 hours pre- and post- procedure. This is a safe and commonly used method of prophylaxis. There is evidence suggesting this does not reduce AKI in patients with GFR > 30 (AMACING, Lancet 2017)
2. N-Acetylcysteine: 1.2g IV q12h x2 pre and x2 post procedure. Theoretically improves renal perfusion and acts as an anti-oxidant. The only high-quality RCT conducted did not find any benefit to the use of NAC (ACT, Circulation 2011). Its use is controversial.
3. Statins: some studies suggest a possible benefit of high-dose statins in reducing CIAKI. This is not a routine performed.
4. Stop nephrotoxic medications: commonly ACEI/ARBs, NSAIDs, diuretics
5. Dialysis does not reduce the risk of CIAKI

NEPHROTIC SYNDROME

Syndrome of substantial urinary protein loss and edema. Characterized by:

1. Nephrotic range proteinuria (> 3.5 g/d)
2. Edema
3. Hypoalbuminemia
4. Hypercoagulability
5. Hyperlipidemia

* Note that patients will often have normal or near-normal creatinine at presentation as this is an issue with permeability rather than clearance although the glomerulus is eventually damaged leading to renal dysfunction.

Etiologies	Clinical Findings/Workup
Glomerular Disease: Nephrotic Syndromes (Proteinuria >3.5g/d) FSGS: 1° vs. 2°: HIV, heroin, pamidronate, steroids, lithium, heme CA, obesity, hyperfiltration MN: 1°: Anti Phospholipase 2 Receptor Antibody vs. 2°: solid organ CA, infection (HepB/C, syphilis), autoimmune (SLE, thyroidits), gold, penicillamine MCD: malignancy (heme, solid organ), infection (HepC, HIV, TB, syphilis, mycoplasma), Drugs (NSAID, antimicrobials, lithium, penicillamine, bisphosphonate, sulfasalazine), SLE Systemic Disease: DM, amyloid, SLE *Primary Renal Pathologies Can be Classified as Primary or Secondary (underlying etiology)	Proteinuria (> 3.5 g/d), hypoalbuminemia, significant edema, hyperlipidemia 24-hr urine collection (Nephrotic proteinuria) Benign urine sediment with fat bodies Renal biopsy Rule out secondary causes (see left column)

HYPERNATREMIA

Serum Na > 145 mmol/L

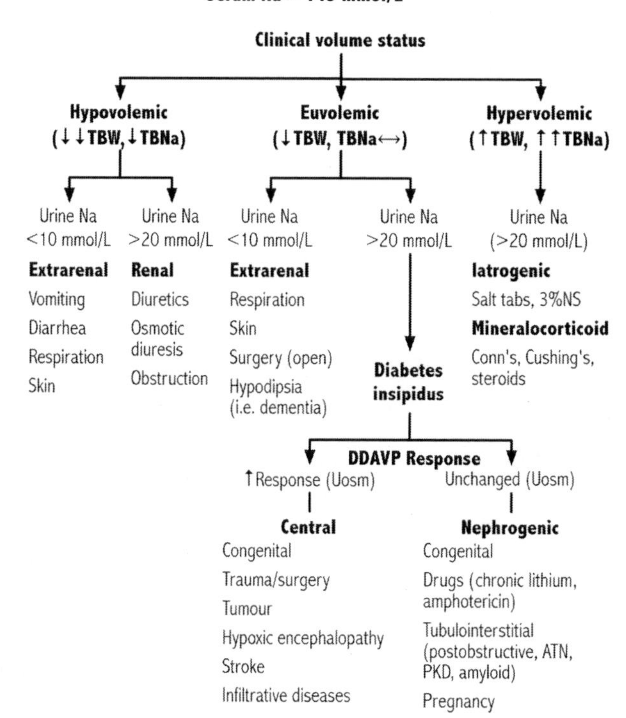

Definition:

- Na >145 mmol/L; Usually a deficit of water relative to sodium.
- Dehydration: hypernatremia due to reduced free water alone (impaired thirst, physically unable to reach water)
- Hypovolemia: hypernatremia from both loss of salt and water (e.g. diarrhea, vomiting)

Clinical manifestations:

- Thirst; N/V; weakness; headache; muscle spasm; neurologic - irritable, delirium, decreased consciousness, coma, seizures
- Fluid depletion: orthostasis, oliguria, history of volume or blood loss

- Diabetes insipidus: polydipsia, polyuria, culprit medication, intracranial insult
- Symptom severity based on magnitude and acuity of onset (more acute = more symptomatic)

Evaluation Summary:
- History: Intake (restricted access, dementia), volume loss (N/V/D), neurologic (stroke, trauma, surgery), Cushingoid features
- Medications: 3% NS, salt tablets, lithium (nephro. DI), diuretics (hypovolemia), steroid use
- Exam: volume status, neurological exam, Cushingoid features
- Investigations: lytes, urea, Cr, glucose, Ca, serum osmolality, urine studies (lytes, osm, urea, cr)

Workup for DI:
Workup for DI if clinically suspected - U_{osm} <300 (complete) or U_{osm} 300-600 (partial) in the setting of hypernatremia
- Water restriction test: Used in cases of polyuria and DI without hypernatremia. Urine volume and osmolality q1h, plasma sodium and osmolality q2h. Stop test if: 1) Uosm >600 (normal – primary polydipsia); 2) Stable Uosm over 2-3h despite rising Posm; 3) POsm > 300 or Na > 145. In 2) and 3), inability to effectively concentrate urine despite rising Posm will diagnose DI.
- DDAVP test – central vs nephrogenic: 10 mg intranasally or 2 ug SC. Measure U_{osm} and volume q30 min x2h.

Diagnosis	U_{osm} after dehydration (mOsm/kg)	Δ U_{osm} after vasopressin (% increase)
Normal	>700	<10%
Central DI		
Complete	<300	>50%
Partial	300-700	10-50%
Nephrogenic DI		
Complete	<300	<10%
Partial	300-500	10-50%
Primary polydipsia	>500	<10%

Treatment Approach:
1. Treatment of underlying cause is crucial
2. If patient is hypovolemic, replace initial fluid deficit with crystalloids first. This will slow correction of hypernatremia.
3. Provide free water or other hypotonic fluids (see chart) to replace remaining free water deficit (see below)
4. Monitor sodium levels q4h initially for crystalloid/D5W titration – do not overcorrect

Correction Recommendations:
- Acute Hypernatremia (<48h): Correct to normal or near-normal in <24h. Give DDAVP if due to DI.

- Chronic Hypernatremia (most cases): Correct <10meq/L/d, or <0.5meq/L/h.
- ******Note if unsure, need to assume chronic for safety******
- Overcorrection: can rarely lead to cerebral edema. Slow your fluid rate.

Fluid Management:

1. This formula does not account for iso-osmotic fluid deficit coexisting with pure water deficit (i.e. crystalloid replacement) or account for ongoing urinary losses (which may be substantial if problem is diabetes insipidus)
2. Estimate free water deficit (WFD – in L) = ($Na_{current}$ -140)/140 x total body water
   - TBW = Wt in kg * % Water
   - TBW % Water estimate: Young – 50-60%; Elderly= 40-50%; higher in males
3. Determine hourly rate of correction (see above recommended rates)
4. Calculate Δ serum Na from 1L infusate = ($Na_{infusate}$ – Na_{serum})/(TBW +1)
5. Calculate infusate volume required to reduce by desired Na (usually 10pt/24h): Δ serum Na desired/(Δ serum Na/1L infusate)
6. Divide answer from #5 by 24h to attain hourly infusate rate to correct current hypernatremia
7. Add 30-40cc/h for sweat and stool losses. Urinary electrolyte-free water losses will need to be considered if significant output.

Example

- 50M presents with Na =168 mmol/L, weight = 68 kg. Infusion of D5W, correcting 10 mmol/L/24h is planned.
- TBW estimate= 68kg * 0.5 = 34L
- FWD estimate = 168-140/140 * 34L = 6.8L
- Sodium change per 1L of D5W: ($Na_{infusate}$ – Na_{serum})/(TBW +1) = [0–168] ÷ [34+1]= –4.8
- Required volume for 10mmol/L Na change: Δ Na desired/ Δ Na/L = 10/4.8 = 2.1L
- Hourly rate of D5W for 10mmol/L Na reduction in 24h: 2.1L/24h = 88cc/h D5W
- Correction for 30-40cc/h obligatory losses: 120-130cc/h D5W
- Remaining free water required for correction to Na = 140: FWD – 24h D5W given = 6.8L – 2.1L = 4.7L

Management By Etiology:

- **Hypervolemic hypernatremia**: stop sodium sources, replace free water deficit, consider loop diuretics (induces natriuresis)
- **Hypovolemic hypernatremia**: Replace ECF depletion with crystalloid, then correct FWD
- **Central DI**:
  - Correct hypernatremia (usually Na normal or mildly elevated due to intact thirst)
  - Solute restrict (protein, salt) – reduced renal solute load decreases obligate urine output (fixed urine osmolality)
  - Desmopressin 5-20ug nasally QHS titrated to nocturia, or 0.05mg-0.8mg po daily in divided doses
  - Other treatments: thiazides, chlopropamide, carbamazepine, NSAIDs
- **Nephrogenic DI**: Solute restriction, thiazide, amiloride, discontinue offending medications (e.g. Lithium)

Fluid tonicity:

Fluids (1 L)	Osmolality (mosm)	Na content (mmol/L)	Other
D5W	278*	0	278 mmol/L glu *Glu is metabolized D5W acts as free water
0.45% Normal Saline	154	77	77 mmol/L Cl
0.9% Normal Saline	308	154	154 mmol/L Cl
2/3 Dextrose - 1/3 Saline	273	51.3	185 mmol/L glucose
Ringer's lactate	273	130	28 mmol/L lactate, 4 mmol/L K^{+}. 1.5 mmol/L Ca^{2+}
3% Normal Saline	1026	513	513 mmol/L Cl

HYPONATREMIA

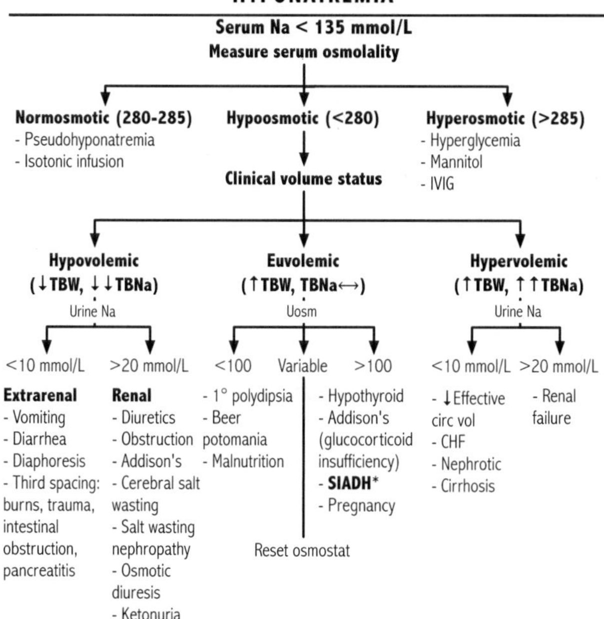

Causes of SIADH*

- Physiologic (stress)
- Malignancy (SCLC, pancreas, duodenal, thymoma, lymphoma)
- Pulmonary – TB, empyema, pneumonia
- CNS: trauma, stroke, hemorrhage
- Drugs: anti-depressants, anti-psychotics, anti-convulsants, many others

In hyperglycemia, corrected [Na] = measured [Na] + 0.3[glucose]
Pseudohyponatremia (now corrected by LAB)→ marked hyperlipidemia, hyperproteinemia (eg. multiple myeloma)

Clinical manifestations:

- Chronic/gradual hyponatremia to 110 mM usually only presents with mild symptoms
- Rapid progression of hyponatremia to 115 mM or less may cause severe symptoms
- Mild symptoms – anorexia, N/V, headache, irritability, falls/instability, confusion, muscle weakness and cramps.
- Severe symptoms - seizures, coma, death

Management principles:

- Treat underlying cause.
- Asymptomatic hyponatremia: correct <0.5 mEq/L/hour. Caution with elderly females, alcoholics, malnourished patients, and burn victims
- Severely symptomatic: initial rapid correction of Na until symptoms resolve (regardless of Na change with treatment)
- Maximum correction of <10mmol/L in 24hrs for all other patients
- Correction for symptomatic or very severely hyponatremic patients; 3% NS (approx. 1mg/kg/h)
- Emergency (symptomatic) treatment: 100mL bolus 3%NS for a rapid increase of 4-6mEq. Consider repeat x1.
- Overcorrection can lead to osmotic demyelination syndrome 2-6d post: dysarthria, dysphagia, paraparesis or quadriparesis, tremor, incontinence, hyperreflexia, cranial nerve palsies, mutism, locked in syndrome, behavioural disturbances, lethargy, and coma)
- If overcorrection occurs, reduction of Na is preferred – D5W, desmopressin can be used
- Fast correction is common in primary polydipsia – this rarely causes ODS as it is almost always acute in nature
- **Hypovolemic** – crystalloid repletion. Na will recover rapidly as ADH stimulus removed – follow Na closely
- **SIADH** – free water restriction (<1.0 L/day) and Tx underlying cause, +/- salt tabs if chronic with no CHF, +/- Lasix.
- **Hypervolemic** – Na and free water restriction (<1L/d.), Lasix
- Options for resistant disease: vasopressin receptor antagonist (i.e. tolvaptan), demeclocycline and oral urea
- Check electrolytes Q2-4h initially to ensure not rapidly correcting – if overcorrect give free water to bring Na+ conc back down

Calculations:

See: Hypernatremia section for TBW determination
Example: 68 F with dehydration, secondary to diarrhea. Weight = 60 kg. Na = 116 mmol/L. To treat with 0.9% NS.

1. Infusate Calculation: Adrogue 2001
2. Change in Na/L infusate = (Infusate Na– serum Na)/(TBW+1)
3. Rate of infusion in mL/h = (desired rate of Na change per hour/change in serum sodium per liter of infusate) * 1000

Change in Na/L infusate = (154 – 116)/(31) = 1.225 Na mEq/L
Rate of infusion (mL/h) = (0.5mEq/h)/(1.225mEq/L) * 1000 = 408mL/h

HYPERKALEMIA

Serum K > 5.0 mmol/L

Clinical manifestations:

- Neuromuscular: weakness, paraesthesias, depressed reflexes, ascending paralysis
- Cardiac: ECG changes → Peaked T wave, flattened P, prolonged PR interval, depressed ST segments, prolonged QRS duration, no P waves, progression to sine waves, VF, VT

Causes:

- **Pseudohyperkalemia**
  - Repeated fist clenching, hemolysis, thrombocytosis, leukocytosis, improper storage of specimen
- **Transcellular shift**
  - Acidosis, DKA, hyperosmolality, non-selective B-blockers, digoxin overdose, hyperkalemic periodic paralysis, succinylcholine, cell lysis (e.g. rhabdomyolysis, tumor lysis syndrome, GI Bleed)
- **Decrease renal K+ excretions**
  - Reduced aldosterone secretion: hyporeninemic hypoaldosteronism (diabetic nephropathy, NSAIDs), all angiotensin inhibitors (ACEi, ARBs, direct renin inhibitors), 1° adrenal insufficiency, chronic heparin
  - Aldosterone resistance: K sparing diuretics (spironolactone, eplerenone, amiloride), certain antibiotics (e.g. trimethoprim), diabetes, SLE
  - Reduced distal solute delivery: decrease ECV (including CHF, cirrhosis, nephrotic syndrome, protein losing enteropathy)
  - Acute/chronic kidney disease (see section on *renal failure*)
- **Increase K+ intake (Food never the sole etiology)**
  - Foods (bananas, potatoes, tomatoes, squash, spinach, milk, cantaloupe etc.)

Approach:

1. Exclude pseudohyperkalemia – blood should be drawn without fist clenching. Hemolysis can be induced. Exclude leukocytosis/thrombocytosis – draw blood by ABG syringe and hand-deliver to the lab. **Repeat lytes if hemolyzed.**
2. Determine if patient is symptomatic or arrhythmia present. If yes, then consider stat:
   - Membrane stabilization: IV Calcium Gluconate 1g IV (10cc, or 1 ampule "amp"); can repeat q5min if concerning ECG changes not improving
   - Potassium shift into cells: 1 amp D50W IV then Humlin R 10 units IV – temporary measure; 1 amp HCO_3
   - Telemetry if arrhythmias or ECG abnormalities (Peaked precordial Tw, loss of P waves, wide QRS)
   - Repeat electrolytes q2-4h prn
   - Long-term potassium removal as below
3. For patients with K>5.0, consider long-term potassium removal:
   - Dietary modifications: excessive intake rarely leads to hyperkalemia on its own but can increase potassium in the setting of other non-modifiable contributing factors

- Fluids and Diuresis: Loop and thiazide diuretics increase urinary loss. Do not diurese if hypovolemic.
- GI cation exchangers: Kayexelate is commonly used but may not be more effective than laxatives.
- Dialysis: if K+ remains elevated despite medical therapy, especially in the context of CKD

4. Medications: hold or stop medications that may lead to hyperkalemia – ACEI/ARB, MRAs, K supplements, NSAIDs
5. Determine and reverse underlying etiology

HYPOKALEMIA

Serum K < 3.5 mmol/L

Exclude: redistribution/cellular shift or spurious hypokalemia (eg. metabolic alkalosis, insulin, increased catecholamines, theophylline, hypokalemic periodic paralysis)
Note that the majority of potassium is intracellular so small changes in serum (extracellular) potassium represent larger total body potassium deficiencies

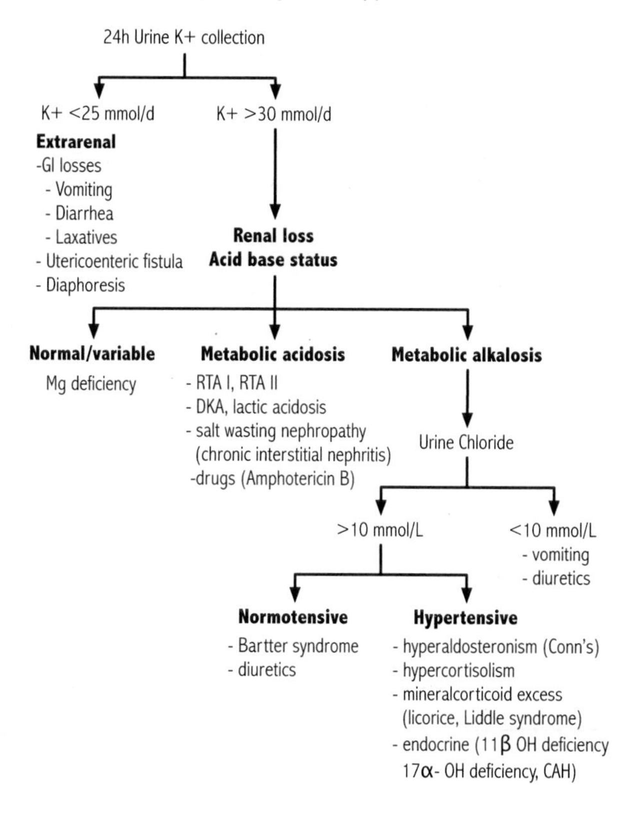

Clinical features:

- **Neuromuscular**: weakness, fatigue, paralysis, ileus, depressed DTR's, parasthesias, respiratory muscle dysfunction.
- **Cardiac**: arrhythmias (sinus bradycardia, paroxysmal atrial or junctional tachycardia, AV block, VT, VF) ECG: flat T's, U waves, ST depression
- **Renal**: poor concentrating ability (nocturia, polydipsia, polyuria), increased renal bicarb reabsorption, increased renal ammonia production, nephrogenic DI.

Management:

- If ECG instability or digoxin toxicity: cardiac monitoring with telemetry
- Replacement: Give PO or NG replacement if possible
  - Oral: KCL elixir 20 mmol/15 ml or K-lyte (K citrate) 25 mmol/packet, Slow K 8 mmol/tab, Micro K 8 mmol/tab
    - Aim for approx. 50-100 mmol per 1 mmol K increase. **More mmol replacement is required as the K level becomes lower.**
  - Maximum IV concentration:
    - **Central line (with cardiac monitoring)**: KCl 20-40 mmol/L (higher rates risk arrhythmias)
    - **Peripheral**: Up to 10mmol/h
    - Higher peripheral concentrations lead to pain due to severe phlebitis
- Replace magnesium deficiency, can lead to refractory hypokalemia
- Treat underlying cause

Potassium:

*Note: Total deficit is often quoted as approx. 100mEq per 1mmol/L; however, correlation between mEq and K change is highly variable, and thus no such corrective formula should be definitely used in practice

Replacement regimen:

- Mild-Mod (K 3.0-3.5):K-elixir 40mEq x1-2, or K-lyte 50mEq x1-2, or slowK/Micro K 2 tablets BID x2-4 (1tab = 8 mEq)
- Severe (K <3): K-elixir 40mEq or K-lyte 50mEq TID to QID, repeat lytes again in several hours. Add IV KCl – if running fluids, IVNS + 40KCl/L. If not running fluids, KCl 10mEq IV x3 (each takes 1h finish)
- Concurrent hypophosphatemia: order KPhos 22/15 mEq IV x1-2, if severe
- Concurrent hypomagnesemia: 2g IV, 5g IV, or PO therapy

Magnesium:

Amount:

- 1g of IV MgSO4 has 8mEq Mg. MgOxide 400mg has 20mEq Mg. MgRougier has 0.4 mEq/mL.
- * Most magnesium stores are intracellular, leading to drop in levels after replacement.
- * Consider PO magnesium for ongoing magnesium losses (**Note: Oral Mg can cause diarrhea, leading to worsening hypomagnesemia)

IV Replacement:

- Mild-Mod (Mg 0.4-0.66): MgSO4 2g IV x1-2
- Severe (Mg < 0.4): MgSO4 5g IV x1

PO Replacement (Mild or Chronic):

- Magnesium Oxide: 400 mg 1-2 tabs daily
- Magnesium Rougier (Liquid): 1-2 tablespoons (15-30mL) 1-3 times daily with meals

Phosphate:

Formulations:

- Phosphate Novartis 500mg tablets (PO): 16mmol PO4 per tablet
- Potassium Phosphate (IV): KPhos 22/15 (i.e. 15 mmol PO4)
- Sodium Phosphate (IV): NaPhos 20/15 mEq (i.e. 15 mmol PO4)

PO Replacement:

- Mild (PO4 > 0.4-0.5): 1 mmol/kg elemental/d, 2-4 doses daily; e.g. phosphate Novartis 2 tabs x1 or x2
- Mod-Sev (PO4 < 0.4-0.5): 1.3mmol/kg elemental/d, 2-4 doses daily; e.g. phosphate Novartis 2tabs x3-4 doses, BID-QID
- *Reduce dose by 50% if renal dysfunction

IV Replacement:

- Mild (PO4 > 0.4): 0.1-0.25 mmol/kg elemental over 6 hours; e.g. NaPhos 20/15 or KPhos 22/15 IV x1
- Moderate-severe (PO4 < 0.4): 0.25-0.50 mmol/kg elemental over 8-12 hours; e.g. NaPhos 20/15 or KPhos 22/15 IV x2

NEUROLOGY

COMA AND IMPAIRED LEVEL OF CONSCIOUSNESS

Initial Management

ENSURE ABCs. Vital Signs. GCS. O2 to keep sats > 90%. Cardiac monitor. NPO.

STAT large bore IV NS.

CBG, ABG, CBC, lytes, extended lytes, albumin, Cr, liver enzymes, lactate, coags, B12, TSH, serum osmolality.

Stat capillary glucose.

BC x2, UA, LP (in correct clinical context) ECG, and chest xray.

Empiric IV Bolus: thiamine 100 mg IV (PRIOR TO DEXTROSE), D50W 50cc IV, Narcan 0.4 µg

Establishment of code status is crucial. Until then give supportive care, treat underlying causes and consider ICU transfer.

Diagnostic Approach

CNS: Reticular Activating System (brainstem, thalamus, hypothalamus, cerebral cortex), large stroke with midline shift

Metabolic: Na, K, Ca, Mg, glucose, renal, hepatic, thyroid, acid-base, hypercarbia, hypoxemia, hypothermia, hypothyroidism, Addisonian crisis

Drugs: sedatives, anticholinergics, anticonvulsants, opioids, Lithium, MAOs, salicylates, amphetamines

Infection: meningitis, encephalitis, abscess, sepsis

Toxins: methanol, ethylene glycol, heavy metals, carbon monoxide, ethanol

Structural: ischemia, intracranial bleed, trauma, tumour, increased intracranial pressure

Psychiatric: catatonia

Seizure: non-convulsive status epilepticus, post-ictal state

Clinical Assessment:

- Collateral history: paramedics, relatives, neighbours, old charts, medical alert bracelets
- For inpatients: has the patient had a recent fall? Are they on anticoagulation?
- Level of consciousness: **GLASGOW COMA SCALE (3-15) / 15**

FEATURE	1	2	3	4	5	6
Eye Opening	None	Pain	Speech	Spontaneous	XXX	XXX
Verbal Response	None	Sounds	Words	Confused	Oriented	XXX
Motor Response	None	Extension	Flexion	Withdraws	Localizes	Obeys

- Complete physical exam for underlying and acute medical conditions.
  - Substance abuse: track marks, stigmata of chronic liver disease
  - Trauma: seat belt sign, blood or fluid in ear, CSF rhinorrhea, raccoon eyes
  - Infection: meningismus, petechial rash
- Eyes: pupil size, reaction and symmetry
  - Pinpoint pupils - pontine lesions, opioids
  - Dilated pupils - anticholinergics, TCAs, sympathomimetics
  - Asymmetric pupils – herniation, CN3 palsy

  - Fundi - Visualize both discs, look for venous pulsations
- CNS: neuro exam – look for localizing signs (asymmetric signs suggest structural lesion or hypoglycemia)

Investigations: (*guided by clinical assessment*)
- Should consider CT head and LP in any case of altered level of consciousness NYD
- Infection without raised ICP: lumbar puncture - send for stat gram stain, glucose, pH, protein, cell count & differential, virology (HSV2, VZV), consider also HIV, syphilis, mycology in appropriate settings. See "Empiric Therapy by Diagnosis" section for treatment of CNS infections.

Subsequent management is based on ***underlying etiology.*** *See appropriate Red Book sections.*

ETOH WITHDRAWL AND DELIRIUM TREMENS

Alcohol Withdrawal Syndromes

Symptoms: at least 2 of:
- autonomic hyperactivity, hand tremor, insomnia, nausea/vomiting, transient hallucinations, psychomotor agitation, anxiety, GTC seizures

Minor withdrawal	Tremulous, palpitations, diaphoresis	6-36 hours
Seizure	Generalized	6-48 hours
Alcoholic hallucinosis	Visual, auditory hallucinations, no autonomic instability	12-48 hours
Delirium tremens	Delirium, + alcohol withdrawal	48-96 hours

Diagnostic and management principles

1. Secure ABCs and rule out mimics
- O2, cardiac monitor, IV access
- DT is a clinical diagnosis and of exclusion of other diagnosis– need full history/physical to rule out infection, trauma, metabolic, hepatic failure, overdose
- Labs: cap blood sugar, lytes, phosphate, calcium, albumin, magnesium, CBC, LFT, toxicology, CK, Cr, ethanol level ± ABG if indicated.
- Imaging: consider CT head if clinically appropriate.
- Initial therapy: rehydration with IV NS, thiamine 300mg IV then glucose (D50W) 50 cc to prevent Wernicke's encephalopathy or Korsakoff's syndrome. Continue thiamine 100mg IV/PO daily indefinitely and appropriate electrolyte and nutritional replacement.

2. Give BENZODIAZEPINES (long-acting preferred)
- CIWA protocol order sets available at both HHS and St. Joseph's
  - Allows RN to perform routine assessment and administer benzodiazepines based on symptoms.
- Diazepam 5-10mg IV q5-10 minutes until initial symptomatic control then q4-6h PRN
- Lorazepam 2-4mg IV q20 minutes until symptomatic control then q1-2h PRN in patients with advanced cirrhosis
- Dilantin is not an effective treatment in withdrawal seizures; use benzodiazepines

3. PHENOBARBITAL if refractory DTs after discussion with SMR / staff
- Phenobarbital 130-260mg IV q20 minutes if refractory to high-dose benzodiazepines (>50mg diazepam or >10mg lorazepam in first hour required for symptomatic control)
- If phenobarbital required will need ICU admission for intense monitoring and often intubation for management of hypotension and respiratory suppression.
- Avoid anti-psychotics as this reduces the seizure threshold. If needed, use low dose Haldol.

SEIZURE AND STATUS EPILEPTICUS

Definitions:
- Seizure – can be focal with or without impairment of consciousness, generalized (either primarily or secondarily)
- Status epilepticus - any seizure lasting > 5 mins OR more than one seizure without recovery in between. Can be *convulsive* or *non-convulsive* (non-convulsive status epilepticus is difficult to identify clinically).

Etiology of seizure: "STATUS SEIZE" – treat underlying cause!

Seizure medication: non-compliance, epilepsy refractory to treatment
Tumor of CNS (rare)
Anoxia/hypoxia (rare)
Trauma of head
Uncontrolled blood glucose (hypo/hyper)
Stroke (rare)
Substance abuse (withdrawal: EtOH, benzos, barbiturates)
EtOH (withdrawal/abuse) (see "DELIRIUM TREMENS")
Infection (meningitis, encephalitis, sepsis)
Zany medications (drugs lowering sz threshold: INH, cephalosporin, penicillin, quinolones, TCAs, bupropion, lithium)
Electrolyte/metabolic abnormalities

Diagnostic and management principles:

1. Secure ABCs and give empiric therapy
- Oral/nasal airway, 100% O2, cardiac monitor, ensure IV access
- Labs: CBG, ABG/VBG with lactate, lytes, extended lytes, albumin, CBC, drug levels, toxicology, CK
- Benzo + initial therapy
  - **Alcoholic**: IV NS, thiamine 300 mg IV then glucose (D50W) 50 cc if known alcoholic
  - **Pregnancy**: Magnesium sulfate 4-6 g IV load then 1-2 g/hr IV
  - **INH toxicity**: 5 g pyridoxine IV
- Collateral: as much possible detail from witnesses (prodrome, head or eye deviation) but *do not delay treatment*
- Rapid exam: vitals, head trauma, fever, meningismus, intoxication, antiepileptic drug compliance, focal CNS findings
- Low threshold to call RACE or code blue if additional help is required or concerned about airway protection

2. Give BENZODIAZEPINES for active convulsions lasting > 5 mins
- Lorazepam (0.1 mg/kg): 2-4 mg IV q3-5 min to max dose of 0.1mg/kg. Maximum infusion rate of 2 mg/minute.
- Diazepam (0.1mg/kg): 2-10 mg IV q5 min to max dose of 20 mg/dose
- If no IV, may give diazepam 0.5 mg/kg PR or midazolam 5-10 mg IM, 5mg buccal

3. PHENYTOIN (will prevent recurrence)
- IV load: 20 mg/kg in 250 mL NS (run at 25 to 50 mg/min) – this is often *underdosed*, estimate weight
  - Can give simultaneously with benzodiazepines in 2nd IV line
  - Run in NS (dextrose will precipitate the drug).
  - Monitored setting required. Watch for hypotension, bradycardia → can slow infusion, but give full dose.

4. If seizures continue:

- **Early call to ICU/SMR** if seizure >5-10min for possible **ICU Admission, Intubation**
- Low threshold to call RACE or code blue if additional help is required or concerned about airway protection
- **Phenobarbital**: 5mg/kg IV over 10 min q10 mins until max dose of 10-20 mg/kg reached. Slow rate if seizures stop, but give full dose.
- **Midazolam** infusion: 0.2-0.5mg/kg/hr (50mg in 50mL NS),
- **Propofol** infusion: 1-2mg/kg bolus, then 0.3-3mg/kg/hr (500mg in 50mL)

Epileptogenic seizure vs. Psychogenic non-epileptogenic seizures:

- Can be difficult to distinguish between the two but very important due to potential harm of treatments for prolonged seizures
- Features of psychogenic seizure: side to side head turning, asynchronous movements, awareness/responsiveness during event despite bilateral motor activity, pelvic thrusting, tongue biting is rare

SYNCOPE

Definition: Transient loss of consciousness and postural tone secondary to cerebral hypoperfusion with NO residual neurological deficits.
Differentiate from: Migraine, seizure, cerebrovascular event

CAUSE	CHARACTERISTICS	EXAMPLES
CARDIAC (~ 23%)		
Arrhythmia (tachy or brady)	Sudden syncope with no warning symptoms, syncope while lying down or sitting, palpitations	Bradycardia: AV block (conduction system disease and/or medications); sinus pauses/brady (vagal, sick sinus, negative chronotropes); pacemaker malfunction Tachycardia: VT/VF (structural heart disease; MI); SVT
Structural	Syncope on exertion, chest pain, dyspnea	HOCM/IHSS, aortic stenosis, aortic dissection, PE, MI, tamponade

NON-CARDIAC (~ 59%)		
Reflex mechanisms (Neurocardiogenic = vasodepressor or cardioinhibitory)	Warmth, nausea, tunnel vision, decreased hearing , lightheadedness, sometimes a specific trigger	Vasovagal, post-micturition/defecation, swallowing, cough
Carotid sinus hypersensitivity	Triggered by neck pressure or head turning	Carotid sinus syndrome
Orthostatic hypotension	Triggered by position change, signs of dehydration or autonomic dysfunction (diabetes, neuropathy, etc.)	Dysautonomias (diabetes, spinal cord injury), fluid depletion, illness, drugs (antidepressants, sympathetic blockers)
Psychogenic (diagnosis of exclusion!)	Frequent episodes	Anxiety, panic
SYNCOPE OF UNKNOWN ORIGIN (~ 18%)		

Diagnostic Approach to Syncope:
The patient's story often gives you the diagnosis!
Rule out: hypoglycemia, MI, PE, and arrhythmia

1. **History**: Associated symptoms (chest pain, dyspnea, palpitations, headache), positional changes (orthostasis), presyncopal prodrome (lightheadedness, tunnel vision, decreased hearing) anxiety/stressors, hyperventilation (parasthesias, cold extremities), triggers (fasting, meals, coughing, micturition), relationship to exercise (effort syncope may be AS, post-effort syncope may be HOCM/IHSS). Medications, past medical history (cardiac disease, stroke, anemia, seizures, diabetes, neuropathies), family history (syncope, sudden death)

2. **Physical**: Special attention to postural vitals, blood pressure (both arms), heart murmurs, neurological deficits.

3. **Investigations:** CBC, lytes, Cr, CBG, trop, extended lytes, albumin, liver enzymes, TSH, B12, fasting lipids. ECG and CXR

   At risk for heart disease? Consider echo, exercise test, telemetry/holter monitor. If holter is nondiagnostic, consider electrophysiologic studies, loop ECG monitor
   Other: consider carotid dopplers, EEG, psychiatric evaluation

Consider risk of serious outcome:
San Francisco syncope rule

- Systolic BP <90 mmHg
- Patient c/o SOB
- Hx of CHF
- Hct <30%
- Abnormal EKG (any non-sinus rhythm or any new changes)

≥1 predicts serious outcome at 30 days, including death, arrhythmia, MI, PE, stroke, SAH, need for transfusion, hospital admission → sensitivity 98%; specificity 56%; If score of 0, safe to discharge home with investigation as an outpatient

Differential Diagnosis of Stroke

	Mechanism	Etiology
Ischemic	Atherosclerotic plaque/ in-situ thrombosis or artery to artery embolism	Aortic arch, Carotid or intracranial disease, small vessel disease (lacunar stroke),
	Cardiac embolism	Atrial fibrillation, valvular disease or prosthesis, LV thrombus
	Other less common causes	Hypercoagulable states (APLA), dissection, vasculitis
Hemorrhagic	Intracerebral hemorrhage	Hypertension, cerebral amyloidosis, AVM, cerebral vein and sinus thrombosis
	Subarachnoid hemorrhage	Aneurysm
Mimics	Intoxication, infections, metabolic (renal failure, hypoglycemia, hyponatremia, hepatic encephalopathy), migraines, syncope, seizures (Todd's paralysis, post-ictal state), structural (trauma, tumour, bleed), psychiatric (conversion disorder)	

History:
- Last seen normal (LSN) time is *essential* and will guide eventual management.
- Get best collateral history possible from patient, caregivers, witnesses
- PMHx: stroke risk factors including hypertension, atrial fibrillation, diabetes, atherosclerosis, smoking

Investigations:
- Urgent non-contrast CT to rule out hemorrhage (sensitivity for ischemic stroke 50% in first 6 hours); consider repeat imaging at 24-48h if first CT negative. MRI optimal if suspected lesion in posterior fossa. It is not possible to differentiate ischemic from haemorrhagic stroke on clinical grounds.
- Labs: CBC, Lytes, Creatinine, INR, PTT, glucose, HbA1c, fasting lipid profile, LFTs, other tests such as hypercoagulable or vasculitis screen based on the clinical presentation;
- Carotid dopplers for anterior circulation ischemic stroke or CT/MR angiography; echocardiography (TTE or TEE +/- bubble study); ECG daily x3, Holter monitor or longer term monitoring for atrial fibrillation.

Management Approach (see HHS pre-printed order set for acute ischemic stroke):

- Assess airway, breathing and circulation; O2 and IV
- Based on LSN time:
    - If <4.5h, consider thrombolysis with tPA if no absolute contraindications +/- endovascular treatment, activate acute stroke team
    - If < 6h, consider endovascular treatment, activate acute stroke team
    - **If inpatient at hospital, contact acute stroke team via Criticall (1-800-668-4357)**
    - Review inclusion and exclusion criteria as part of HHS acute thrombolysis order set online
- Vitals and neurovitals q4h x 24h and then reassess
- Physical examination and neurological assessment using *NIH stroke scale*
    - Standardized assessment used by healthcare providers to evaluate stroke severity

BP	For patients not receiving tPA: Should not be lowered unless SBP>220, or DBP>120 or hypertensive emergency Permissive hypertension x48hrs and until neurologically stable
Blood Sugars	Avoid extreme hyperglycemia; maintain glucose at about 10 (avoid hypoglycemia)
Antiplatelet	ASA 160mgx1, then 81mg daily Consider clopidogrel (75mg daily) if patient allergic to ASA.
Anticoagulation	• VKA or NOAC should be initiated for stroke secondary to A. Fib after 3-5 days for small infarctions, 6-10 days for medium size infarction 9-15 days for large size infarctions, after repeat CT head to rule out hemorrhagic transformation • UFH in the acute setting is of no proven benefit, avoid unless recommended by neurology for special indications.
Treat vascular risks	Statin, blood sugar control
Preventative	• NPO until swallowing assessment is completed– prevent aspiration, SLP consult • DVT prophylaxis (LMWH) • Bladder scans / in/out catheterizations; avoid foley catheters • Bowel routine • Early mobilization / multidisciplinary team management

- Monitor for complications:
    - Neurologic deterioration: recurrent stroke, cerebral edema, hemorrhagic transformation, seizures
    - Other: MI, arrhythmia, aspiration and pneumonia, dehydration, urinary retention and infection, DVT/PE, malnutrition, pressure sores, contractures

Transient Ischemic Attack (TIA):

Definition: transient episode of neurologic dysfunction caused by focal brain/spinal cord/retinal artery ischemia without infarction. Note that TIA is a tissue/imaging diagnosis, not based on time (as it was previously)

Investigation: Non contrast CT head, ECG, fasting lipids, HbA1c, carotid dopplers for anterior circulation TIA (all performed on an urgent basis: within 24-48hrs)

Management:

- Cardioembolic TIAs (AFib) – anticoagulation (VKA, NOAC)
- Non-cardioembolic TIA - antiplatelet therapy with ASA or clopidogrel (if allergy)
- Carotid endarterectomy in symptomatic carotid stenosis:
  - Carotid stenosis > 70-99% and life expectancy > 5 yrs: NNT to prevent one stroke over five years = 6.3
  - Carotid stenosis 50-69% and life expectancy > 5 yrs: NNT to prevent one stroke over five years = 22
  - No benefit for total/near total occlusion at five years
  - No benefit for less than < 50% stenosis
  - Earlier is better, ideally within 2 weeks
- Aggressive risk factor modification: treat DM, lipids, HTN, smoking cessation, reduce alcohol intake
- Urgent referral to the Stroke Prevention Clinic

OBSTETRIC / GYNECOLOGICAL EMERGENCIES

HEMODYNAMIC ADAPTATIONS IN PREGNANCY

Cardiac	Blood volume ↑ 50-60%, peaks at end 26-30 weeks Cardiac output (↑ SV) 50% ↑ HR 10-20 (normal <105) BP ↓T1 (decreased SVR), nadirs 10mmHg below baseline by 20-24weeks and increases T3 normalizing by term BP usually falls immediately after delivery and then tends to rise reaching a peak postpartum day 3-5	BP <140/90
Resp	Progesterone ↑respiratory drive (↑ RR) Minute ventilation ↑ + tidal volume ↑, ↓ERV Compensated respiratory alkalosis	PcO_2 >30 HCO_3 >22-24 O_2 Sat >97%
Renal	↑renal plasma flow (60-80%) + increased GFR = Drop in Cr Progesterone stimulates RAS → 50% increased blood volume	Cr 30-70
GI	High HCG causes nausea, vomiting max T1 ↓motility and delayed gastric emptying = constipation, GERD ↑ ALP due to placental production ↑ Symptomatic gallstones = Bile is more lithogenic and saturated with cholesterol	No change AST/ALT/Bili Alk Phos <220
Heme	Hypercoagulability up to 6-8 weeks post partum Dilutional anemia and thrombocytopenia can occur Neutrophilia	DDimer + 30% T1>90% T3 Hb >110, WBC <13 No change INR
Endo	Maternal insulin resistance to shunt glucose preferentially to the fetus BHCG is a weak thyroid stimulator can see subclinical hyperthyroidism T1	TSH 0.2-2.5 T1 Use T4 (not T3)

OB HISTORY

Think pregnancy related and non-pregnancy related causes for all presentations!

Usual history **PLUS**:

- Prior pregnancies (G- total pregnancies, P- preterm, T- term, A- abortion)
- Current gestation (T1 1-13wks, T2 13-26wks, T3 26-40wks), >24 weeks –fundus at the umbilicus *viable fetus
- Past complications (eg. Gestational HTN, Preeclampsia, Gestational Diabetes, PE)
- Prenatal screening (HIV, Rubella, HepBSAg, VDRL, Toxo, CMV, TB, Varicella, gonorrhoea/chlamydia)
- Rupture of membranes, bleeding, contractions, fetal movement, last fetal monitoring (CTG)

Approach to Pre-Existing Medical Conditions in Pregnancy:

- How does pregnancy affect the disease
- How does the disease affect pregnancy
- What are the medication considerations
- What are the implications in the peripartum period
- What would I do if the patient was not pregnant

OB CODES

Standard of Care for Pre-Code Situations:

- If >20 weeks notify obstetrics team (if not already present), low threshold to review with staff
- Place patient in left lateral decubitus or tilt to decrease IVC compression
- Give 100% oxygen and ensure experienced airway operator present (if warranted)
- Establish IV access above the diaphragm
- Consider causes

CAUSES ("**BEAU-CHOPS**")

- **B**leeding/DIC
- **E**mbolism (PE, Coronary, amniotic fluid)
- **A**naesthetic complications
- **U**terine atony (post-partum)
- **C**ardiac disease (MI, ischemia, aortic dissection, CM)
- **H**ypertension/Preeclampsia/Eclampsia
- **O**ther: DDx of standard ACLS guidelines
- **P**lacenta abruption/previa
- **S**epsis

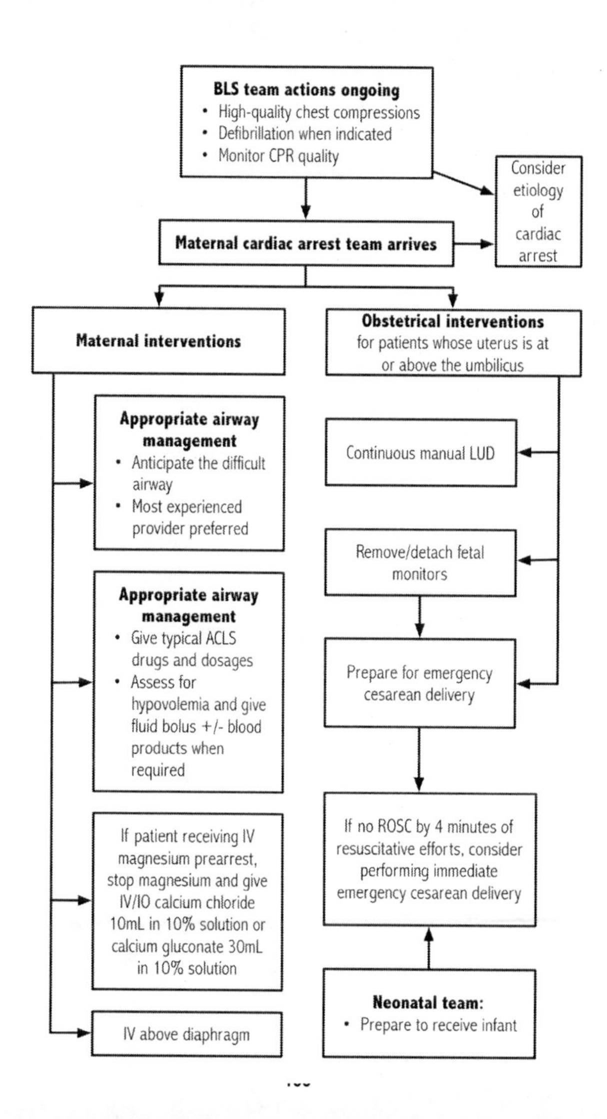

BLS team actions ongoing
• High-quality chest compressions
• Defibrillation when indicated
• Monitor CPR quality
Consider etiology of cardiac arrest
Maternal cardiac arrest team arrives
Maternal interventions
Obstetrical interventions for patients whose uterus is at or above the umbilicus
Appropriate airway management
• Anticipate the difficult airway
• Most experienced provider preferred
Continuous manual LUD
Remove/detach fetal monitors
Appropriate airway management
• Give typical ACLS drugs and dosages
• Assess for hypovolemia and give fluid bolus +/- blood products when required
Prepare for emergency cesarean delivery
If patient receiving IV magnesium prearrest, stop magnesium and give IV/IO calcium chloride 10mL in 10% solution or calcium gluconate 30mL in 10% solution
If no ROSC by 4 minutes of resuscitative efforts, consider performing immediate emergency cesarean delivery
IV above diaphragm
Neonatal team:
• Prepare to receive infant

HYPERTENSIVE DISORDERS IN PREGNANCY

Definition: HTN in Pregnancy is defined as >140/90 with severe HTN >160/110

Classification:

1. **Pre-existing (Chronic) HTN:** occurs pre-pregnancy or <20 weeks gestation
2. **Gestational HTN:** occurs > 20 weeks gestation
3. **Preeclampsia:** systemic syndrome (occurring after 20 weeks gestation) that causes maternal endothelial injury; leading to vasoconstriction, capillary leak, and activation of the coagulation cascade. (TIP: Think about it like SEPSIS). It is heterogeneous in timing, symptoms and clinical presentation. It is a progressive illness that will not resolve until the placenta is delivered. Timing of delivery is an obstetrical decision

Diagnosis:

- Pre-existing HTN – with one or more of :
    - 1) resistant HTN >20 weeks despite treatment
    - 2) new or worsening proteinuria,
    - 3) presence of adverse finding(s) OR severe complication(s).
- Gestational HTN with at least one of:
    - New proteinuria (At least 1+ on dipstick, verify → >300mg/day or Urine Protein:Creatinine ratio > 30mg/mmol).
    - Presence of adverse finding(s)
    - Severe complication(s)
- **Severe Preeclampsia:** Preeclampsia + severe complication(s) & requires immediate delivery

	Adverse Findings (Increased Risk)	Severe Complications (Needing Immediate Delivery)
Neurologic	Headache, Visual Changes	Seizures, PRESS, Cortical Blindness, Decreased LOC, CVA/TIA
Cardiovascular	Chest Pain	BP 160/110 despite active treatment, MI
Respiratory	SOB, Sat <97%	Pulmonary Edema
Hematologic	Thrombocytopenia	Plts <50, INR >2
Renal	Elevated Cr	AKI Cr >150 , need for dialysis
GI	Nausea, Vomiting, RUQ pain, LFT rise	Hepatic Rupture, HELLP
Fetal/Placental	Abnormal HR, IUGR, Abnormal dopplars	Abruption, Absent or reversed end diastolic flow, fetal demise

- **Post Partum Preclampsia:** preeclampsia that manifests in the post partum period. Usually within 48 hours but can be as late as two weeks post delivery. *Have a high index of suspicion for any women requiring readmission for blood pressure elevations post partum.*

Investigations:
- CBC, Smear, INR, LFT's, Bili, Cr, Albumin, LDH, Fibrinogen
- Urinalysis, Urine PCR or 24 hour urine collection
- ECG, CXR if clinically indicated

BP Medications for Use in Pregnancy and Breastfeeding (*Captopril and Enalapril can be considered if breastfeeding post partum)

	Long Acting	Short Acting (For BP >160/110)
Labetalol	100-400mg po TID	20-80mg IV q30min, then 1-2mg/min infusion
Nifedipine	XL: 20-60mg po BID	SA: PO/SL 5-10 mg q 30 mins
Methyldopa	250-500mg po BID-QID	---
Hydralazine		5-10mg IV q30min, 0.5-10mg/hr

Gestational or Chronic Hypertension Management:
- Target 130-150/80-100mmHg if no comorbidities or <135/85 with comorbidities (diabetes, kidney disease, cardiac disease, LVH)

Preeclampsia Management:
- Definitive management is **delivery** in Severe Preeclampsia, you are there to provide supportive care to the Obstetric Team
- **Supportive measures:** ABC's, O_2, IV access, NPO, LLD, monitor BP, urine output and neurological status
- **Urgent BP Management:**
  - Initiate Short Acting Tx when SBP > 160 or DBP > 110 to reduce maternal risk of stroke, target DBP 90-105
- **Magnesium sulfate:**
  - Recommended as prevention of eclampsia (seizures) in Severe Preeclampsia *and* as treatment for eclampsia
  - Initial loading 4 g IV over 15-20 min then 1-2g/hour unless Cr >150/anuria (then discuss with staff unless seizing)
- For recurrent convulsions, give additional 2 g IV over 5 min.
- In the absence of IV access, may give initial loading dose as deep IM injection 5 g each buttock (total 10g).
- No role for phenytoin or benzodiazepines unless there is a contraindication for MgS04 or is ineffective
- **Monitor reflexes and respiratory rate!** *Toxicity: loss of reflexes, low resp rate (<12 breaths/min), nausea, flushing, headache, lethargy, hypocalcemia, bradycardia, ECG changes, hypotension, flaccid quadriplegia* → ***give calcium gluconate IV 1 amp***

From the SOGC Guidelines 2014

GI ISSUES IN PREGNANCY

Remember:

- Always include your pre-pregnancy differential diagnosis (i.e. Viral Hepatitis and CMV can be fulminant in pregnancy.)
- A palpable liver is abnormal in pregnancy (liver shifts posteriorly and superiorly)

	Hyperemesis Gravidarum (HG)	Intra Hepatic Cholestasis of Pregnancy (IHCP)	Hemolysis Elevated Liver Enzymes Low Platelets (HELLP)	Acute Fatty Liver of Pregnancy (AFLP)
Gestation	T1	T2-T3	>20 weeks	T3 (34-36)
Symptoms	Refractory nausea, vomiting → ketosis, dehydration, weight loss	Itch (often starts on palms and soles) no rash	Asymptomatic → General malaise, RUQ pain, nausea, vomiting	Prodrome of polydipsia, malaise, nausea, vomiting, confusion, RUQ pain,
Sick or Not Sick	Not sick (can be dehydrated)	Not sick	Sickish→ Sick	SICK
AST/ALT	Mild to <500	<1000	1000's	<500
Bilirubin	No	Usually normal	Hemolysis	Severe↑
Other	Ketosis, Hypokalemia	↑Bile acid levels	Thrombocytopenia	BW = acute liver failure Low glucose, *SWANSEA Criteria
Treatment	Supportive management	Ursodiol 12mg/kg divided daily bid or tid Routine NST Delivery before 38 weeks Association with autoimmune and PSC	Deliver *If not resolving 48 hours post partum think about TTP (more common in pregnancy)	Deliver Hepatology Consult Test for LCHAD deficiency in mother/fetus

Approach to dyspnea:

- Dyspnea is common by T3 (80-90%), but hypoxia and/or persistent breathlessness laying flat is NOT.
- TIP: **Walking pregnant patients may unmask hypoxia**
- **Resp**: PE, Amniotic Fluid Embolism Pneumonia/Aspiration, Asthma
- **Cardiac**: Peripartum cardiomyopathy, Tocolytic induced pulmonary edema, Undiagnosed VHD,CAD,CHD, Pre-eclampsia with excessive iatrogenic fluids
- **Metabolic**: Sepsis, or other causes of lactic acidosis
- Disordered breathing of pregnancy
- Mechanical restriction from gravid uterus (more common with polyhydramnios, twins, late gestation)

Amniotic Fluid Embolus:

- Caused by amniotic fluid in the maternal circulation which likely occurs via trauma, endocervical veins or placental insertion site. Can occur during labour, delivery or post-partum.
- Can cause profound hypoxia due to VQ mismatch and non-cardiogenic pulmonary edema, acute pulmonary HTN and eventual biventricular failure
- Can lead to respiratory failure, DIC, coma, seizures or death
- Hx: SOB, CP, hypoxia, hypotension or shock
- **Diagnosis**: Clinical and by exclusion of all other etiologies
- **Management**:
  - Supportive, rx hypoxemia and hypotension, should be in an ICU setting
  - NE or dopamine are reasonable inotropic agents
  - Cautious IVF as pulmonary edema is common
  - Decision to deliver if occurs intrapartum
  - 10-90% maternal mortality in severe cases!

PE: (see Hematology section for PE/DVT)

- Prediction rules are not validated in a pregnant population. A D-Dimer will be positive in >90% women by T3 and 100% post partum.
- Treatment: Enoxaparin 1.5mg/kg daily or 1mg/kg bid (recommended), IV UFH if acute reversal anticipated DOACs not recommended (no data)

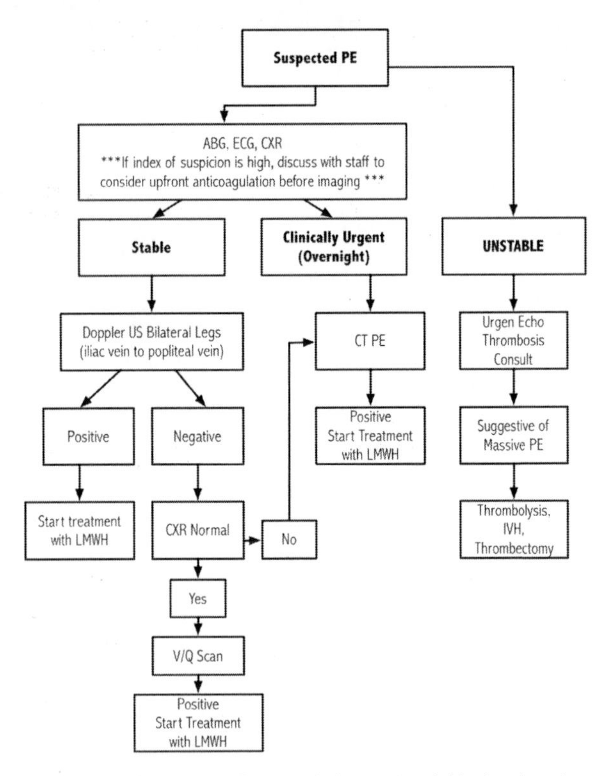

Radiation Doses *No Single Imaging Test exceeds the recommended fetal maximum in pregnancy

	Fetus	Mother-Breast Radiation
Chest CT	0.2	20-25mGy
V/Q	0.9	0.28mGy
CXR(PA + lat)	<0.01	---

CARDIAC DISEASE IN PREGNANCY

Approach to Chest Pain:

- All pregnant patients with chest pain requiring analgesia should have a positive diagnosis before discharge. Don't forget typical cardiac risk factors.
- TIP: A raised resp rate, chest pain, persistent tachycardia and/or orthopnea should always be fully investigated.
- **Resp**: PE, Pneumonia with pleural involvement, pulmonary arterial hypertension
- **Cardiac**: SCAD (coronary dissection), CAD, undiagnosed or worsening congenital or valvular disease, hypertensive emergency, arrhythmia, peri/myocarditis
- **Aortic**: Dissection (history of Connective Tissue Diseases)
- **GI**: Reflux, dyspepsia
- **MSK**: Diagnosis of exclusion but is more common with hormonal changes and compression

Cardiac Disease In Pregnancy:

- Whatever the underlying causes of cardiac insufficiency, the ability to tolerate a pregnancy is related to presence of pulmonary HTN, hemodynamic significance of any abnormality, functional NYHA class, or presence of cyanosis (O2 sat <80%).
- **Any degree of cardiac dysfunction from any reason usually requires a cardiology or obstetric medicine consult (even if asymptomatic).**

Peripartum Cardiomyopathy:

- Definition: EF <45% occurring in the last month of pregnancy or up to 5 months (usually presenting <1month) after giving birth after exclusion of other causes.
- Risk factors: multiple pregnancy, hypertension in pregnancy, multiparity, advanced maternal age, previous PPCM, preeclampsia
- Etiology unknown.
- Many patients recover completely, however some are left with a progressive dilated CM
- Sx: SOB, decreased exercise tolerance, palpitations, pulmonary edema, orthopnea, cough
- Symptomatic Mgmt: O2, diuretics, initiate beta-blockers in those with systolic dysfunction once stable, consider afterload reduction with hydralazine and nitrates. Expedited delivery once out of acute CHF and prompt initiation of ACE and other disease modifying agents post partum.
    - Usualy involves input from cardiology, OB, anaesthesia/cardiac anaesthesia in decisions around timing and mode of delivery

DRUGS IN PREGNANCY

Need to have informed conversation with patient.
Resources: Reprotox (Free for Trainees) and Lactmed (Breastfeeding)

Drugs that are usually contraindicated:

Drugs	Safer Alternative
ACE / ARB	Labetalol, Nifedipine XL, Hydralazine, Methyldopa
Warfarin	LMWH/UFH
Chemotherapy	Patients with breast cancer may be treated with AC after T2
Antibiotics: tetracyclines, aminoglycosides, TMS-SMX, fluoroquinolones	Cephalosporins, Penicillins, Ertapenem or Meropenem, Clindamycin, Nitrofurantoin
AED: Phenytoin, Valproic Acid, Carbemazepine, Topiramate (unless no other AED controls seizures, risk higher if multiple different AED used)	Lamotrigine, levetiracetam (limited evidence)

ONCOLOGY

FEBRILE NEUTROPENIA

Definition:

- Temperature > 38.3C or > 38.0C for more than 1 hour associated with absolute neutrophil count (ANC) < 0.5 cells/mm^3, or < 1.0 cells/mm^3 if expected to decrease to less than 0.5 cells/mm^3
- Focus of infection identified in only 20-30% of patients

History

- Malignancy and specifics of chemotherapy treatment (hx of mucositis, problems with CVCs)
- ID History: farms, animals exposure, sexual history, travel, TB; any antibiotic prophylaxis
- Other associated symptoms (ex. Organ specific), ROS
- Are there other non-infectious causes for fever (uncontrolled cancer, blood transfusion, PE)

Physical:

- Signs of inflammation may be minimal or absent in the severely neutropenic patient
- Vital signs including postural changes
- Examine central venous catheters (PICC, port), skin, H&N (teeth, pharynx, sinuses, fundi), chest, abdomen (include inspection of perianal area, but ***no DRE***), feet
  - o RLQ abdo tenderness may be suggestive of typhlitis (neutropenic enterocolitis- life threatening); may require further imaging

Investigations:

- **Empiric antibiotics AFTER blood cultures are drawn, and ***before any other investigations*****
  - o Cultures: BC from every intravenous port and at least one peripheral; Urine R+M, C+S
- CBC with differential, lytes, Cr, AST, ALT, GGT, ALP, Bili, INR, aPTT, lactate
- CXR, AXR → CT Chest/abdo/pelvis
- In correct clinical context:
  - o Respiratory Symptoms: NPS
  - o Diarrhea: Stool for C. difficile, norovirus
  - o Catheter appears infected: Swabs for C+S of site
  - o Neuro changes: Lumbar puncture
  - o Neutropenia >7d: consider serum galactomannan
- If persistent fever, consider other occult sources

Management:

- Empiric antibiotics AFTER blood cultures are drawn; goal is to cover most likely/most virulent pathogens
- Pre-printed febrile neutropenia protocol available through HHS
- Involve hematology/medical oncology if known to either service, and infectious disease

- Review previous culture data (i.e. need to cover ESBL or MRSA if previously grown)

No defined focus	Beta lactam with anti pseudomonal coverage	Tazocin 4.5g IV q6h ***OR*** ceftazidime 2g IV q8h
	Penicillin Allergic	Ciprofloxacin + metronidazole + vancomycin
Suspect MRSA infection	Skin/soft tissue infection, mucositis, ? catheter related infection, + MRSA previously, hemodynamically unstable, on fluoroquinolone prophylaxis	Add vancomycin 1gram q12h
Suspect Fungal infection	Neutropenia >7d Hemodynamically unstable/febrile 4-7 days after initiation of broad spectrum antibiotics	Add voriconazole, caspofungin

- High risk patient: expected to be neutropenic for >7d, hemodynamically unstable, hematological malignancy, comorbidities, hepatic or renal dysfunction, mucositis, higher intensity chemotherapy, prolonged neutropenia, inpatient at time of FN dx, altered mental status
- Consider GCSF support in high risk patient (discuss with Heme/Med Onc)
- Can consider outpatient management in the appropriately selected low-risk patient with oral ciprofloxacin and amoxicillin-clavulanate
  - MASCC score can be used to risk stratify (Klastersky et al. J Clin Oncol. 18:3038-3051. 2000.)

HYPERCALCEMIA OF MALIGNANCY

Mechanism:

Osteolytic metastasis	Breast, Multiple Myeloma, Lymphoma, Leukemia
Tumor section of PTH related protein (PTHrP)	Squamous cell (H&N, esophagus, cervix), Carcinoma (breast, ovarian, renal, bladder), Lymphoma, Leukemia
Extra renal production of Vitamin D	Lymphoma, ovarian dysgerminomas
Ectopic PTH secretion	SCLC, ovarian, thyroid, papillary tumor, pancreatic, rhabdomyosarcoma, neuroectodermal tumor

Clinical manifestations:

- Neuro: weakness, fatigue, personality change, delirium → decreased LOC → coma
- Cardiac: short QT, wide QRS, bradycardia, conduction blocks
- Renal: DI, AKI, nephrolithiasis, nephrocalcinosis, distal RTA, polyuria
- GI: nausea, vomiting, anorexia, constipation, pancreatitis
- Pain: bony (symptom or cause?)
- Psych: anxiety, depression, cognitive dysfunction

Decision to treat:

- Consider the degree of calcium elevation, rate of increase and severity of symptoms

Mild <3	Encourage fluid intake (If symptoms → IVF) Avoid: thiazides, lithium, dehydration, bed rest, calcium in diet
Moderate 3-3.5	IVF + Bisphosphonate
Severe >3.5	IVF + Calcitonin rescue + Bisphosphonate (+ Hemodialysis ex. if CHF, AKI)

Acute Treatment:

IVF	Promote renal clearance of Ca^{2+}	200-300mL/hr Maintain 100-150mL/hr of urine output	Consider cardiac status. May need to tx resulting volume overload with loop diuretic
Bisphosphonate	Bone builder = Ca^{2+} deposits in bone	Zoledronic Acid (4mg IV) Pamidronate (30-90mg based on Ca^{2+} level IV)	**Effect = 4-7 days** Toxicity: flu like symptoms, nephrotoxicity, ON of the jaw
Calcitonin	Bone building, renal Ca^{2+} clearance, ↓ GI reabsorption	4IU/kg subcut q12h	Onset 1-4hrs Tachyphylaxis: effective for 48hrs

Chronic Treatment:

- Poor prognostic sign: 50% of patients die within 30 days of onset
- Definitive treatment = disease control +/- routine bisphosphonate

SPINAL CORD COMPRESSION

Definition:

- Compression of thecal sac and contents (spinal cord, cauda equina) by an extradural tumor mass
- Diagnosis: clinical and radiologic (minimum indentation of the theca at level of clinical symptoms)
  - Thoracic (60%), lumbosacral (30%), cervical (10%)
- Prostate, multiple myeloma, lymphoma > breast, lung, renal cell carcinoma
  - Consider non-malignant causes: epidural abscess, hematoma, disc herniation, spinal stenosis, bony metastases with no cord compression, radiation myelopathy
- Can rapidly lead to paralysis, even in metastatic disease it is important to treat to prevent permanent neurologic deficit in order to maintain quality of life **therefore have a high index of suspicion**

Mechanism:

- Vertebral metastasis → invasion of epidural space → obstruction of venous plexus → vasogenic edema → cord infarction
- Bony metastases leading to mechanically unstable vertebrae and compression of cord

History:

- Pain (90%, often first sign): severe localized, progressive back pain, worse recumbent, radicular pain.
- Weakness (60-85%): symmetric → loss of gait function (walking, stairs) → paralysis
- Hyperreflexia below the lesion, if lesion is in cauda equina, depressed deep tendon reflexes in the legs
- Sensory disturbance: ascending numbness and paresthesias
- Sphincter Dysfunction (late): urinary retention → overflow incontinence, bowel incontinence

Physical exam:

- Palpate the spine for bony tenderness
- Neurologic examination
  - Strength: symmetrical weakness
  - Sensory: level typically occurs 1-5 levels below the compression
  - Reflexes: UMN or LMN pattern depending if cord or cauda equina being compressed
  - DRE to assess for loss of rectal tone
  - Gait: ataxia

Investigations:

- MRI spinal cord protocol (can be multiple lesions, image entire spine) – if on call, contact radiologist overnight. If cannot be done, test should be prioritized in AM.
- CT myelogram if MRI not an option (similar sensitivity and specificity)

Management:

- Goals: pain control, avoid complications, preserve or improve neurologic function

ACUTE	Treat vasogenic edema	Dexamethasone 10mg IV, then 4mg po/IV q6h
	Treat pain	
DEFINITIVE	Neurosurgery	Posterior decompression + laminectomy, +/- Stabilize spine, tissue diagnosis if initial presentation
	Radiation	Pain and local tumor control

- Urgent consultation to Radiation Oncology and Neurosurgery
- All cases warrant a formal neurosurgery opinion, but must consider how many spinal levels are involved, patient's overall status and prognosis to determine who is an appropriate surgical candidate

Prognosis:

- Determined by the neurologic function at the time cord compression is diagnosed (try to diagnose early!)
- Median survival 6/12 post presentation
- Better: ambulatory at presentation (8-10/12), breast, prostate, lymphoma, multiple myeloma
- Worse: non ambulatory at presentation (2-4/12), lung

TUMOUR LYSIS SYNDROME (TLS)

Definition:

- Massive and abrupt release of cellular contents (nucleic acids, phosphate, potassium) into bloodstream after rapid lysis of malignant cells
- Treatment induced: 12-72hrs post initiation of cytotoxic treatment for hematologic malignancy; solid tumors = rare, but at risk if large, highly proliferative
- Spontaneous: rare, highly proliferative tumors with high disease burden (Burkitts)

↑ K^+	≥6.0 mmol/L or 25% ↑ from baseline	Cardiac Arrhythmia, Heart Failure
↑ Uric acid	≥476 mmol/L or 25% ↑ from baseline	AKI, Gout
↑ PO_3^{4-}	≥1.45 mmol/L or 25% ↑ from baseline	Nausea, vomiting, diarrhea, lethargy
↓ Ca^{2+} (2° to ↑ PO_3^{4-})	≤1.75 mmol/L or 25% from baseline	Tetany, cardiac arrhythmia, seizure * Chvostek/Trousseau
AKI	Secondary to ↑ Uric acid and Calcium phosphate deposition in kidneys. Impairing clearance of K, PO_3^{4-} and uric acid.	

Management:

- Cardiac monitoring for arrhythmia, urine output
- Lytes/Creatinine/uric acid level monitoring q4-6h
- Consider early involvement of Nephrology re: dialysis

↑ K^+	Stabilize cardiac membranes: calcium gluconate 1 amp (if ECG changes) Shift: 1 amp D50 + 10 units insulin R, ventolin Remove: Kayexylate
↑ Uric acid	Rasburicase 0.2mg/kg IV daily x5/7
↑ PO_3^{4-}	IVF, Refractory: phosphate binders
↓ Ca^{2+}	Treat if symptomatic (tetany, arrhythmia), otherwise, do not treat until phosphate corrected to minimize calcium phosphate precipitation
AKI	Aggressive IVF +/- diuresis to maintain urine out put Indications for dialysis: persistent hyperkalemia, volume overload, uremic symptoms, symptomatic hypocalcemia, severe oliguria or anuria

Prevention:

- High risk patients: haematological malignancy (Burkitt, AML, Diffuse Large B cell), solid tumors (high tumor burden, germ cell tumors), CKD, LDH >2x upper limit, older patients
- Prophylaxis prior to chemotherapy (high risk patients): IV hydration, allopurinol (prevents further production) or rasburicase (acutely lowers uric acid by metabolizing it)

SUPERIOR VENA CAVA SYNDROME

Definition:
- Obstruction of venous return from the head, neck and upper extremities due to external compression or intrinsic obstruction of the SVC causing elevated central venous pressures
- Lung cancer, lymphoma, mediastinal tumours
- Thrombosis of central venous catheter, post radiation fibrosis, infections (TB, blastomycosis, histoplasmosis), substernal goiter, LAD

Clinical Manifestations:
- Swelling/edema of face, neck, upper extremities, dyspnea (orthopnea), cough, stridor (laryngeal edema), hoarseness, dysphagia; headache, confusion, ischemia (cerebral edema)
- Cough, chest pain, arm swelling
- Distended collateral veins on the face/neck/chest, H&N exam for lymphadenopathy
- Typically symptoms peak and then gradually improve as venous collaterals form

Investigations:
- CXR – may identify lung tumour or mediastinal widening → CT chest with contrast

Management:

ACUTE – decrease venous congestion	Elevated head of bed Limit fluids, consider diuretics Corticosteroids (if steroid responsive tumor – lymphoma) – Dexamethasone 10mg IV, then 4mg po/IV q6h
ACUTE – Crisis (laryngeal edema, cerebal edema)	Protect airway → intubate, possibly corticosteroids Endovascular stent Thrombosis → anticoagulate, ?thrombolysis
DEFINITIVE	Chemotherapy, radiation therapy Rarely, surgical venous bypass Anticoagulation

- Usually onset of symptoms is gradual and there is time to develop a targeted plan
- If first presentation, ideally obtain tissue biopsy before starting any treatment

LEUKOSTASIS

Definition:

- Elevated blast cell count (>100x10⁹/L- leukocytosis) AND symptoms of decreased tissue perfusion (respiratory or neurological distress)
  - AML: symptoms often occur when WBC >100x10⁹/L
  - CLL: symptoms often occur when WBC >400 x10⁹/L
- Is a medical emergency

Mechanism:

- White cells occlude microvasculature impeding blood flow
- High blast mitotic activity = local hypoxemia, cytokine production → endothelial damage, hemorrhage

Symptoms/Investigations:

Fever	Inflammation due to leukostasis ?Underlying infection	Often treat with empiric antibiotics
Neurological	Visual changes, headache, dizziness, gait instability, confusion, tinnitus, somnolence; increased risk intracranial hemorrhage	CT head MR brain Caution with any contrast dye
Respiratory	Dyspnea, hypoxia; low P_AO2; use SaO2	CXR – diffuse alveolar infiltrates
Bleeding	Up to 40% of patients present with DIC Bleeding history important	DIC screen in correct clinical context
Other	Myocardial ischemia, RV overload; worsening renal insufficiency, priapism, acute limb ischemia, bowel infarction	ECG, lytes, Cr, lactate, physical exam

- CBC (manual platelet count), lytes (potassium from heparinized samples), Cr, extended lytes, urate, DIC screen (fibrinogen, D-dimer)

Management:

- Consult hematology
- IVF Hydration, treat DIC if present
- Cytoreduction (chemotherapy is the only treatment that improves survival)
  - Induction chemotherapy with TLS prophylaxis (allopurinol and aggressive fluids)
  - If cannot have chemotherapy: leukapheresis + hydroxyurea + TLS prophylaxis
  - Hydroxyurea alone = only hyperleukocytosis, no symptoms and cannot start chemotherapy
- Do not increase whole blood viscosity, **AVOID**: blood transfusions (if necessary → slow transfusion), diuretics
- Target platelet count of 20,000-30,000/uL

Prognosis:

- 20-40% of patients with leukostasis die within 1 week of presentation

MANAGING TOXICITIES OF IMMUNOTHERAPIES

Concepts:

- Immune checkpoint molecules maintain peripheral tolerance to self
- Tumors can evade immune recognition by presenting molecules identifying it as "self"
- Immune checkpoint inhibitor antibodies can increase T cell responses and turn the immune system against the tumor; this can also cause an unchecked immune response and autoimmune-like inflammatory responses with collateral damage to normal organ systems and tissues
- Immune checkpoint inhibitor antibodies include: PD-1/PD-L1 inhibitors (nivolumab, pembrolizumab, atezolizumab) and CTLA-4 inhibitors (ipilimumab)
- These are now commonly used in metastatic melanoma and lung cancer

General effects:

- Fatigue: mechanism unknown
    - Watch for symptoms of hypothyroidism (known endocrine toxicity of immunotherapies)
- Pyrexia, chills, infusion reactions: secondary to cytokine release
    - Treat with antipyretics, give antihistamine or corticosteroid for infusion reactions

Organ-specific effects:

- Derm: most common!
    - Rashes: maculopapular, papulopustular, Sweet's syndrome, follicular or urticarial dermatitis, SJS/TEN, oral mucositis, gingivitis, sicca symptoms
    - Maculopapular rash: topical/oral corticosteroids and antipruritics
    - If atypical or non-responsive/severe: kidney and liver labs, consider biopsy, serum tryptase and IgE, oral/IV corticosteroids
    - Hold checkpoint inhibitor if severe
- Diarrhea/colitis: increased stool frequency, abdominal pain, colonic inflammation
    - Mild – less than 4 stools per day: colitis diet, monitor closely, rule out other causes
    - If worse (>4 stools/d) or persists >3d, start oral corticosteroids (1mg/kg), rule out infectious cause, hold checkpoint inhibitor and anti-diarrheal meds, image and C-scope as needed
    - Severe/non-responsive: hospitalization, IV corticosteroids and additional immunosuppression (infliximab) as needed
- Endocrine: includes hypophysitis, hypo-/hyperthyroidism, thyroiditis, adrenal insufficiency
    - Assess pituitary/hypothalamus and end-organ axes (ex. LH/FSH, TSH +T4/T3, prolactin, ACTH, 8am cortisol, IGF1)
    - Consider MRI
    - May be transient or permanent, consider endocrine consult, treat the endocrine abnormality with any necessary hormone replacement
    - Because hormones can be replaced, often do not need to stop checkpoint inhibitor
- Hepatic: mostly asymptomatic transaminitis; can have hepatitis with hepatomegaly, periportal edema/LAD
    - R/O other causes of hepatitis, consider metastases; liver/biliary labs, imaging

- o If hepatitis, corticosteroid taper over a minimum of 3 weeks and hold checkpoint inhibitor, +/- mycophenolate mofetil or antithymocyte globulin for additional immune suppression
  - o DO NOT use infliximab (hepatotoxicity)
- Lung (pneumonitis): SOB, cough, fever, CP - CT chest, blood/sputum cultures; consider ID and resp consult,
  - o Manage conservatively (if mild), hold checkpoint inhibitor therapy and monitor closely. If persistent or interferes with ADLs start oral/IV corticosteroids (1mg/kg) and escalate to other immunosuppressive agents (infliximab, cyclophosphamide, mycophenolate mofetil) as needed

  Consider prophylactic abx

Rare effects:
- Neurologic syndromes: Myasthenia Gravis, transverse myelitis, aseptic meningitis, enteric neuropathy, Guillain-Barre syndrome (corticosteroids and neurology consult; consider IVIG or plasmapheresis)
- Ocular toxicity: uveitis (topical corticosteroids, consult ophthalmologist, consider oral corticosteroid)
- Renal toxicity: AIN (pt usually asymptomatic, Cr gradually rising, most improve with corticosteroids)
- Pancreatic toxicity: asymptomatic lipase elevation that can be monitored; rare pancreatitis

Combinations with other anti-cancer therapies:
- Studies of combinations with chemo/radiation therapy, other targeted therapies, other immunotherapies and antiangiogenics are under way to determine if adverse effects are increased, or changed in spectrum

Pre-existing autoimmune or infectious diseases:
- Unclear if checkpoint inhibitors would worsen HIV/HBV/HCV infections (studies under way)
- Autoimmune conditions in pts requiring a certain level of immunosuppression excludes them from clinical trials based on animal models suggesting fatality, thus it is unclear
- Certain pts with autoimmune disease or immunosuppression from organ transplantation have safely been given checkpoint inhibitor drug

PALLIATIVE CARE

GENERAL APPROACH

- Early, frequent, open communication with the patient and family is key to high quality care
- Ensure values, goals of care are clear to patient, family and medical team
- Symptom assessment tools (eg. ESAS) can be used to identify and monitor symptom burden
- Explore psychosocial and spiritual issues and their impact on symptoms
- Provide information on prognosis if patient/family wish
- Prepare patient and family members for anticipated symptoms/course
- Consider early consultation to palliative care service and/or SW and/or chaplaincy
- Involve allied health team in symptom assessment and management

PAIN

General notes:

- Always consider pain mechanism when deciding on management and add appropriate adjuvant treatments if indicated (see below)
- Use short acting opioid formulations to achieve pain control and switch to slow release once stable dosing and no signs of toxicity/side effects
- Consider route of administration:
  - Do not use oral route if: swallowing difficulty, extreme dyspnea, decreased LOC, decreased GI absorption (eg. vomiting, diarrhea) or if patient is near end of life
  - Patients near end of life: Prefer subcut route as opposed to IV
    - Routine subcut meds are given via butterfly (usually in thigh or upper arm)
- For constant pain, provide scheduled dosing q4h with breakthrough dose (BTD) at 50% regular dose q1h prn
- For incident pain (predictable/severe, and short-lived) associated with specific activities or procedures (eg: dressing changes, transfers, mobilization, personal care), can use prn only if no pain at other times. Administer analgesia as per BTD above at least 30 min (po) or 15 min (subcut) or 5 min (IV) prior to activity, and specify this in the orders
- Slow release formulations should not be crushed or chewed and should ideally not be titrated more frequently than every 5-7 days

Suggested starting regimens based on pain severity:

- Mild pain (ESAS 1-3)
  - Acetaminophen 650mg po q4-6h routine
  - If already using opioid, ensure BTD is available prn
- Moderate to severe pain (ESAS $\geq$4) – choose one of:
  - Morphine 2.5-5mg po q4h + 2.5mg po q2h prn (use half this dose or q6h dosing in renal impairment or elderly)
  - Hydromorphone 0.5-1mg po q4h + 0.5mg po q2h prn (use half this dose or q6h dosing in renal impairment or elderly)
  - For subcut delivery of morphine or hydromorphone, reduce doses by 50%

- o Titrate routine dose up daily based on amount of breakthrough used in past 24h – caveat: patients may request opioids for anxiety or insomnia as well as pain and this can confound your daily total calculation
  - o If not gaining control (and no side effects) consider one-time increase in basal dose by 25%. Watch for effect/side effect.
  - o If patients are already on opioids and have no side effects, use their home dose as a starting point and titrate up based on breakthrough use
- Pain crisis (ie. patient is in distress, pain score is 20/10)
  - o Start with Subcut/IV formulation
  - o Morphine 5-10mg Subcut/IV or dilaudid 1-2mg Subcut/IV stat, then repeat q10-15minutes until pain control achieved (LOC should be monitored more closely if patient is opiate naiive)
  - o Once control achieved, change to po or subcut delivery q4h with BTD, use no more than 50% of the total 24hr dose calculated from what was given to settle the crisis
  - o Consider prn midazolam or lorazepam for sedation (careful monitoring of level of consciousness required, ensure family is aware of why this is being used and is in agreement with sedation)

Considerations for opioid medication:

- Start with lowest effective dose, go slow - titrate every 24-48h in order to achieve pain control
- If pain control not achieved, sum total 24h requirement (scheduled + BTD), divide by 6 for new q4h scheduled dose, add BTD of 50% q2h PRN (or 25-50% q1h PRN)
- If pain controlled, switch to slow release formulation: sum total 24h requirement (scheduled + BTD), divide by 2 for new q12h dose. Ensure to add BTD 10% of total daily dose at q4h interval. Need to specify when to start new regimen and stop old one (eg. At 0900h or 2100h)
- If limited by side effects, consider rotating to a different opioid - If rotating: sum total 24h requirement (scheduled + BTD), use equivalency chart to determine 24h requirement of new drug, **Decrease dose of new medication by 30% to account for variation in receptor affinity for different opioids,** then divide by 6 for new q4h scheduled dose, add BTD of 50% q2h (or 25-50% q1h).
- Remember to always add a bowel regimen
- PO:Subcut/IV ratio is 2:1 for all opioids
- Use only one opioid for scheduled and BTD (except with fentanyl patch will need alternative for BTD)
- Use particular caution in elderly or in renal failure patients, and use hydromorphone as a first choice. Consider dosing routine q6h for these patients unless severe pain
- If patient has renal failure use hydromorphone (avoid morphine, codeine, oxycodone as metabolites are active and renally cleared)

Opioid Equivalencies:

Drug	po (mg)	subcut or IV (mg)
Codeine*	100	50
Morphine	10	5
Oxycodone	5-7.5	N/A
Hydromorphone (Dilaudid)	2	1

**6-10 % of population has no effect from codeine as they cannot metabolize it*

Opioid Side-effects:

- **Respiratory depression**: (rare if appropriate initial dose and titration)
  - Impending resp failure (RR<=6) due to narcotic use (not imminent end of life) - give Naloxone (Narcan) 0.4 mg SC/IV, may repeat q 2-3 mins up to 10 mg – consider lower dosing based on whether patient is a chronic opioid user (~0.04-0.1mg/kg)
  - May need to repeat q 1-2 h, or infuse, if using slow release opioids; naloxone can cause withdrawal or pain crisis in chronic pain or palliative patients, therefore the goal is to reduce respiratory depression without reversing analgesia
  - In non-emergent cases (ie RR 6-12) hold next opioid dose and decrease subsequent doses by 50%. Monitor patient.mpending resp failure (RR<=6) due to narcotic use (not imminent end of life) - give Naloxone (Narcan) 0.4 mg SC/IV, may repeat q 2-3 mins up to 10 mg
  - May need to repeat q 1-2 h, or infuse, if using slow release opioids; naloxone can cause withdrawal or pain crisis in chronic pain or palliative patients, therefore the goal is to reduce respiratory depression without reversing analgesia
  - In non-emergent cases (ie RR 6-12) hold next opioid dose and decrease subsequent doses by 50%. Monitor patient.
- Other:
  - **Constipation (most common)** – ensure patient is on bowel protocol and monitor/titrate to effect.
  - **Sedation** – usually only present initially or with rapid increase in dose. Consider lowering routine opioid dose if it persists.
  - **Myoclonus** – possible sign of opioid toxicity. Lower routine opioid dose and monitor for resolution. If severe and distressing to patient can use midazolam 0.5-1mg subcut q4h prn. If persists, consider opioid rotation.
  - **Nausea** – often only present initially. If persists, consider adding anti-emetic agent or opioid rotation.

Adjuvant treatments:

- Neuropathic pain
  - Gabapentin (Neurontin): start 100-300mg po qHS, Titrate q3days: bid, tid, then qid. next can increase dose from 300mg to 600mg then 900mg. Max daily dose 3600mg. Slow titration over 3 weeks. Particular caution required in elderly and renal dosing required in renal failure/impairment
  - Pregabalin (Lyrica): start at 50-75mg po qHS. Titrate up to bid dosing by 150mg/week. Max daily dose 600mg/day.
  - Nortriptyline: 10mg po qhs, can slowly titrate up to max daily dose 150mg; risk of QT prolongation, anticholinergic side effects
- Pain due to **obstructing or impinging tumour** (inflammatory/visceral pain or raised ICP)
  - Dexamethasone 4mg po/subcut/IV q0800h. Can increase up to 8mg po/subcut/IV q0800h and q1200h. Try not to give later than 1600h as can interfere with sleep. Start at total of 16g/day for treatment of spinal cord compression or brain mets. Wean slowly (ie. 2mg q5days) after definitive treatment achieved. Add a PPI and do not use with another NSAID.
- Bone pain
  - Acetaminophen 325mg – 1g po qid, max 4g/d or 2g/d in liver disease

  - Naproxen 125 – 500mg po bid (with PPI) or Ibuprofen 200-400mg po bid to qid (with PPI) or Ketorolac 30mg IV q8h prn, max 120mg/d. use with caution in elderly or renal patient, not for prolonged use (with PPI)
  - Consider bisphosphonate (eg. Zoledronic Acid, Pamidronate) and/or calcitonin if malignant bony disease
  - Consider Radiation Oncology referral for palliative radiation.
- Muscle spasm
  - Baclofen 10mg po BID can titrate to max dose of 20mg po qid (can cause constipation and delirium)

NAUSEA

Common symptom in advanced disease, causing significant distress. Often multifactorial.

Differential Diagnosis:
- Medications: chemotherapy, opiods, NSAIDs
- CNS: raised ICP (brain metastasis, leptomeningeal involvement, post radiation)
- GI: esophagitis (candida), gastroparesis, gastric outlet obstruction by tumor, constipation, bowel obstruction, radiation treatment involving GI system, ileus, omental metastasis, extensive liver involvement
- Metabolic: hypercalcemia, hyponatremia, renal failure, AI

Management:
- Investigate to determine possible causes
- Treat underlying cause where possible: constipation relief, raised ICP, bowel obstruction
- Non-pharmacologic:
  - Minimize triggers
  - Discontinue offending medications when possible
  - Behavioural interventions for anticipatory N&V
  - Some evidence for complementary medicine techniques – acupuncture, ginger, guided imagery
- Pharmacologic:
  - Choose anti-emetic based on etiology
    - Eg. Dysmotility/gastroparesis – use prokinetic; vestibular disturbance – use anticholinergic/antihistamine
  - Consider the side effect profile of anti-emetics
    - May need to combine different anti-emetics if refractory or due to multiple causes
  - Consider route of administration
    - Haloperidol can be given subcut in small volumes which is easy to tolerate
    - Ondansetron, Gravol and Metoclopramide can be given subcut but in large volumes which is painful for patients

Anti-emetic	Dose	Potential side effects
Anti-psychotic/Dopamine antagonists (for opioid induced):		
Haloperidol (Haldol)	0.5-1 mg (po/subcut/IV) bid-tid (or q8h if occurs at night)	Extrapyramidal reaction (rigidity, dystonia, tremor)
Prochlorperazine (Stemetil)	5-10 mg (po/subcut/IV) bid-qid (q6h if occurs at night) max 40 mg/day	Anticholinergic reaction (dry mouth, constipation, urinary retention, confusion) Extrapyramidal reaction.
Prokinetic (for gastroparesis/dysmotility):		
Metoclopramide (Maxeran)*	10 mg (po/subcut/IV) qid ac meals & qhs (or q6h if nausea is not meal related)	Extrapyramidal reaction, QT prolongation, diarrhea.
Domperidone (Motilium) *	10mg po tid ac meals	
Anticholinergics/Antihistamines (for vestibular/motion induced):		
Dimenhydrinate (Gravol)	25-50 mg (po/pr/subcut/IV) q4-6h prn (12.5 mg in elderly)	Drowsiness, dry mouth, constipation, confusion (particularly in elderly therefore avoid if possible)
Betahistine (Serc)	8-16 mg po tid with meals	
Other:		
Ondansetron (Zofran) **	4-8 mg (po/IV) q8h max 8 mg/day in liver disease	Headache, ***constipation***, QT prolongation
Dexamethasone (decadron)	2-4mg po/subcut/iv od – bid (q0800h and q1200h)	Anxiety, insomnia, hyperglycemia. steroid induced myopathy or adrenal insufficiency with prolonged use.

* *contraindicated in mechanical obstruction, perforation.*

** *Not covered by ODB or most private drug plans.*

DYSPNEA

Subjective experience of breathing discomfort – must ask about it, no reliable physical exam findings

Assessment:

- Inquire about severity, functional impact and level of distress
- Identify and treat exacerbants of dyspnea – eg. Pneumonia, COPD exacerbation, pulmonary edema, pleural effusions

Management:

- Non-pharmacologic:
  - Environmental modification – eg. Access to open window, ***fan***
  - Psychosocial techniques: relaxation strategies, psychosocial support
  - Activity modification: mobility aids, modification of ADL management (aka energy conservation strategies)
  - Chest physiotherapy for mobilizing secretions, respiratory rehab
- Oxygen Therapy:
  - Does not appear to improve dyspnea in non-hypoxemic patients
  - Some evidence for trial of oxygen for symptom relief in hypoxemic patients
- Pharmacologic:
  - Scheduled narcotic can help relieve dyspnea; suggested starting regimens (subcut route is preferred with moderate-severe dyspnea)
    - Morphine 2.5-5mg po q4h + 2.5mg po q1h prn
    - OR Morphine 2.5mg subcut q4h + 1mg subcut q1h prn
    - OR hydromorphone 0.5 – 1mg po q4h + 0.5mg po q1h prn
    - OR hydromorphone 0.25 – 0.5mg subcut q4h + 0.25mg subcut q1h prn
  - See PAIN section for general notes on narcotic use and titration rules
  - If dyspnea not adequately controlled and there is a component of anxiety, consider adding:
    - Benzodiazepam: Midazolam 0.5 – 1mg subcut q4h prn or lorazepam 0.5-1mg po/SL q4h prn
    - Phenothiazine: methotrimeprazine (nozinan): 6.25mg subcut qHS and titrate up slowly
  - Adjunctive therapy:
    - Consider bronchodilators and diuretics as indicated by cause
    - Consider Radiation Oncology referral if indicated (eg. Obstructing tumor)
    - Indwelling pleural drain: recurrent pleural effusions
    - Lymphangitic carcinomatosis: Dexamethasone 4-8mg po/subcut/iv od-bid (q0800h and q1200h)

Diarrhea (non-bacterial):

- Consider and treat potential causes (ie. Meds, decreased GI absorption, pancreatic insufficiency, diet, ***overflow***)
- Loperamide (Imodium) 4mg po x1 then 2mg po after each loose BM
- OR diphenoxylate (Lomotil) 2.5 – 5mg po tid-qid (max 20mg/d)
- Carefully monitor hydration status

Constipation:

- *Non-pharmacologic measures*: optimize intake of fluid, reassess constipating medications (opioids, anticholinergics, iron, calcium), encourage mobility, ensure access to commode, correct electrolytes
- If appropriate, DRE to assess for impaction
- Consider AXR to assess fecal load if uncertain bowel history
- *Pharmacologic* starting regimen:
  - Stool softener:
    - Osmotic Agent: PEG 3350 17 g daily or Lactulose 30 mL po daily-TID
    - Laxative: Senokot 1-2 tabs po qhs or bid (up to 8 tabs/d)
- If no BM by 48h
  - Lactulose 30ml po daily or bid (can produce significant gas and/or cramping)
  - OR **Polyethylene Glycol 3350 17g per day (pt must be able to tolerate 250cc of liquid orally)**
  - OR Milk of magnesia 30-60ml po qhs (caution in renal failure patients)
- If no BM by 72h
  - Rule out impaction
  - Bisacodyl supp 10mg pr
  - Fleet enema (caution in CKD patients) / soap suds / milk and molasses pr if supp ineffective
- If impacted
  - Glycerine supp pr, then disimpaction, then enema
  - Ensure adequate BTD of analgesia given prior to disimpaction
- If refractory
  - AXR to assess fecal loading
  - PEG 3350 17g po tid
  - Relistor (methylnaltrexone) if opioid related – currently difficult to obtain due to manufacturer back-order

Communication:

- Consult palliative care team if symptoms poorly controlled or patient/family/staff needing support
- Communicate with family regarding some or all of the following depending on their wishes: prognosis, goals of care (ie. Recommendations to stop all/some of the below list), expected signs/symptoms at end of life, symptom management strategies, concerns/fears, spiritual needs
- Provide status update and discuss management plan with bedside nurse
- Discuss with Charge Nurse transfer to palliative suite or private room if available, facilitate transfer to hospice or home when possible
- Consider referral for pet visit with Zachary's Paws for Healing if at HHS
- Offer access to chaplain services (or priest) and social work

Communication:

- Stop use of continuous cardiac monitoring
- If pt has an ICD, ensure it is deactivated (if overnight – get magnet from code blue cart/CCU and tape over ICD to deactivate, and call arrhythmia in AM)
- Stop blood work and other investigations (including accuchecks)
- Stop medications other than those for symptom management
- NPO/hold PO meds if decreased LOC
- Stop routine HR and BP assessment. Temp and Resp rate can indicate discomfort so can continue PRN
- Where appropriate, remove IV lines
- Foley catheter can be left in or removed – determine based on individual patient (ie. Leave in if perineal sores/wounds/irritation or if significant pain with care, d/c if causing discomfort, pt delirious, etc)

Medication orders (to be considered):

- Pain:
  - Continue prior ***analgesia*** in subcut route (dose calculated as equianalgesic) . Ensure breakthrough dose available – typically morphine or hydromorphone
  - **Tylenol** 650mg PR q4-6h prn (for fever/pain)
- Secretions:
  - **Scopolamine** (0.4mg subcut q4h prn) can be used to help dry *oral* secretions; (coarse, rattly, snoring/gurgling sounds) as this can be distressing to family members in the last hours of life. Inform family that efficacy of this medication is variable and that excess respiratory secretions are not distressing to the patient – use only if patient actively dying, as if patient more alert, may precipitate delirium
- Agitation:
  - **Haldol** 0.5-1mg sc q4-6h prn (for nausea/confusion/agitation)
  - **Midazolam** 0.5-1mg sc q2-4h prn (for anxiety/restlessness)
- Dry Eyes/Mouth:
  - Oral balance gel and Moi-Stir spray qid and q4h prn
  - **Natural tears** qid routine if sleeping with eyes open, otherwise q4h prn

MEDICAL ASSISTANCE IN DYING

Definition:

- circumstances where a medical practitioner, at an individual's request: (a) administers a substance that causes an individual's death; or (b) prescribes a substance for an individual to self-administer to cause their own death

Criteria (adapted from CPSO): the patient must

1. Be eligible for publicly funded health services in Canada
2. Be at least 18 years of age and capable of making health decisions
3. Have a grievous and irremediable medical condition ➔serious and incurable illness, disease, disability; advanced state of irreversible decline in capability; condition causes enduring physical or psychological suffering intolerable to them and not relieved under conditions they consider acceptable; natural death has become reasonably foreseeable
4. Make a voluntary request for MAID that is not the result of external pressure
5. Provide informed consent to receive MAID after having been informed of the means that are available to relieve their suffering, including palliative care

Process:

- Requires a written request by the patient and assessment and agreement by 2 independent practitioners
- A 10-day period of reflection (can be shortened if death is imminent)
- The patient must be kept informed throughout the process of alternatives and option to rescind
- The prescription must reflect that the substance is intended for MAID
- The prescriber is responsible for completing the medical certificate of death and informing the Coroner
- All MAID deaths are subject to examination by the Coroner

Conscientious Objection:

- Legislation does not compel physicians to provide MAID
- Physicians should communicate their decision not to provide MAID to the patient, provide the patient with all available information and alternatives to address their clinical needs, and make an effective referral to a practitioner willing to provide MAID.

CLINICAL PHARMACOLOGY & TOXICOLOGY

DRUG OVERDOSE AND POISONING

General Principles

1. ABC's; consider indications for airway support, think about IV access and fluid resuscitation, move into monitored setting and consider vital signs requirements

2. Determine specific intoxicant(s) and identify other metabolic derangements:
   - history from patient or collateral history if possible
   - appropriate investigations; (CBC, Lytes, BUN, Cr, Glu, VBG)
   - acetaminophen level, salicylate level, ethanol level
   - serum osmolality and calculate both osmolar gap and anion gap; if suspicious of toxic alcohol can order methanol and ethylene glycol levels
   - consider urine toxin screen, urinalysis
   - 12-lead ECG, CXR
   - beta-hcg (if appropriate)

3. Decontamination:
   - activated charcoal; if within 1-2h of ingestion (longer for some drugs with enterohepatic recycling or delayed gastric emptying, eg, TCAs)
     - not useful for: alcohols, metals (iron, lithium)
     - requires intubation if at risk of aspiration
   - whole bowel irrigation; used for iron pill overdseo otherwise rarely used but consider if sustained-release or enteric coated formulation;
   - hemodialysis in specific situations (i.e. salicylates, methanol, ethylene glycol, lithium)
   - gastric lavage; rarely if ever indicated; unclear benefits with risks of serious complications

4. Antidotes/specific treatment
   - for comatose patients, give "universal cocktail"; glucose, naloxone, thiamine
   - call **Poison Control (1-800-268-9017)** for guidance
   - acetaminophen - NAC
   - ASA - bicarbonate infusion, dialysis
   - beta blocker - atropine, glucagon, high dose insulin
   - toxic alcohols - fomepizole/ethanol and dialysis
   - opioids - naloxone

5. Other considerations
   - identify other metabolic and medical complications
   - concurrent issues in intoxicated patients: AKI, rhabdomyolysis, infections, electrolyte disturbances, intracranial hemorrhage, iatrogenic complications
   - consider involuntary admission if indicated (forms 1 and 42); assess for suicidal intent

Mental Status	Pupils	Vital Signs	Other Sx	Examples
		Sympathomimetic		
Hyperalert, agitation, hallucinations, paranoia	**Mydr-iasis**	↑Temp ↑HR ↑BP ↑Pulse Pressure ↑RR	**Diaphoresis**, tremors, hyperreflexia, seizures, rhabdo, MI/arrest	Cocaine, amphetamines, pseudo/ ephedrine, theophylline, caffeine
		Anticholinergic		
Hypervigilance, agitation, hallucinations, delirium with mumbling speech, coma	**Mydr-iasis**	↑Temp ↑HR ↑BP ↑RR	**Dry flushed skin**/ mucous membranes, ↓ bowel sounds, urinary retention, myoclonus, choreoathe-tosis, picking behavior, seizures(rare)	Antihistamines, TCA, cyclobenzaprine, antiparkinson agents, antispasmodics, phenothiazines, atropine, scopolamine, belladonna alkaloids (eg. Jimson Weed)
		Cholinergic		
Confusion, coma	**Miosis**	↑Temp ↓HR ↑↓BP ↓↑**RR**	**S**alivation, **L**acrimation, **U**rinary/fecal incontinence, **D**iaphoresis **G**I –diarrhea **E**mesis. Bronchocon-striction, bronchorrhea, fasciculations and weakness, seizures	Organophosphates/ insecticides, nerve agents, nicotine, pilocarpine, physostigmine, edrophonium,
		Opioid		
CNS depression, coma	**Miosis**	↓Temp ↓HR ↓BP ↓**RR**	Hyporeflexia, respiratory arrest, needle marks	Heroin, morphine, methadone, oxycodone, hydromorphone, fentanyl
		Sedative-hypnotic		
CNS depression, confusion, stupor, coma	Miosis (usually)	↓Temp ↓HR ↓BP ↓RR	Hyporeflexia Respiratory arrest	Benzodiazepine receptor agonists, barbiturates, alcohols,

Hallucinogenic				
Hallucinations, perceptual distortions, depersonalization, synesthesia, agitation	Mydr-iasis (usually)	↑Temp ↑HR ↑BP ↑RR	Nystagmus	Phencyclidine, LSD, mescaline, psilocybin, designer amphetamines (eg. MDMA, MDEA)
Serotonin Syndrome				
Confusion, agitation, coma	Mydr-iasis	↑Temp ↑HR ↑BP ↑RR	Tremor, myoclonus, hyperreflexia, clonus, diaphoresis, flushing, trismus, rigidity, diarrhea	MAOI, SSRIs, meperidine, dextromethorphan, TCAs, amphetamines
Neuroleptic Malignant Syndrome				
Mental status change, rigidity, fever, dysautonomia	Normal	↑Temp ↑HR ↑BP ↑RR	Lead pipe rigidity, mental status change NOT typical: shivering, myoclonus, hyperreflexia, ataxia, shivering	Haldol, flupenazine, chloropromazine, clozapine, risperidone, metoclopramide
Tricyclic Antidepressant				
Confusion, agitation, coma	Mydr-iasis	↑Temp ↑HR ↓BP ↓RR	Seizures, myoclonus, choreoathe-tosis, cardiac arrhythmias and conduction disturbances	Amitriptyline, nortriptyline, imipramine, clomipramine, desipramine, doxepin

ACETAMINOPHEN

Therapeutic max dose 4g/day (although short term higher dose [7g/day] has been shown to be safe)
Toxicity occurs with ingestions >150mg/kg (fatal and nonfatal hepatic necrosis)

Clinical Manifestations:
- **Stage 1 (<24hrs):** Anorexia, N/V, diaphoresis, pallor, lethargy, malaise, normal labs
- **Stage 2 (24-72hrs):** Stage 1 symptoms improves, increasing AST/ALT (usually 24-36hrs), RUQ pain, hepatomegaly, elevated INR & bilirubin, renal impairment
- **Stage 3 (72-96hrs):** Liver enzyme and function abnormalities peak, Stage 1 symptoms reappear with jaundice, confusion, AKI, death secondary to multi-organ failure
- **Stage 4 (>5days):** Clinical resolution of hepatotoxicity or progression to multi-organ failure

Investigations:
- Acetaminophen level, VBG, BUN, CBC, lytes, Cr, glucose, INR, AST, ALT, bilirubin, lipase, urinalysis, blood & urine tox screen
- Draw acetaminophen level at 4 hrs post-ingestion (or immediately if >4hrs at presentation), and again 2-4 hours later (to determine the trend)
- Repeat acetaminophen level and liver enzymes at 24 hrs post-ingestion if NAC initiated
- Follow nomogram on next page (use SI units)

Management:
1. IV access, oxygen as needed, monitored setting
2. GI decontamination if appropriate
3. **N-acetylcysteine (NAC)**
    - Prevents serious hepatotoxicity and death if given within 8-10 hrs of ingestion
    - PO protocol: 140mg/kg PO load, then 70mg/kg PO Q4H for a total of 17 doses
    - **IV protocol: 150mg/kg IV load over 15-60 min, then 50mg/kg over 4 hrs, then 100mg/kg over 16 hrs**
    - IV protocol is preferred if vomiting ,contraindication to oral administration (pancreatitis),patient preference, presence of liver failure.
    - duration: measure ALT, acetaminophen, and INR when approaching end of protocol; if acetaminophen present or ALT and/or INR are abnormal, continue with 6.25mg/kg/hr IV and reassess with q12h acetaminophen, ALT, and INR measurements
    - can discontinue NAC when acetaminophen undetectable, ALT normal/clearly decreasing, and INR <2.0
    - Always call **Poison Control (1-800-268-9017)** to discuss the details of the case.

General Considerations:

- **Do NOT delay giving 1st dose of NAC while awaiting levels** if there is history of >150mg/kg ingestion or pre-existing liver disease
- May give 1st dose up to 24hrs after ingestion if acetaminophen present
- If hepatic failure develops, continue NAC at 6.25mg/kg/hr until resolution of encephalopathy and INR<2.0, or until liver transplant
- If pre-existing liver disease exist (ie. alcoholics) the indication for NAC should be at lower levels of acetaminophen toxicity

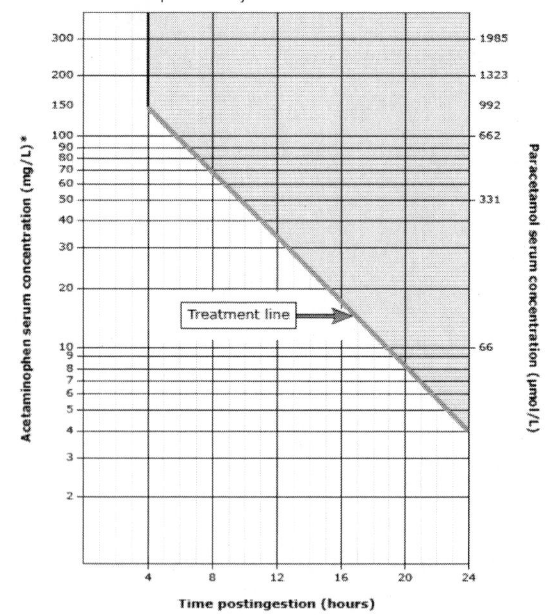

This nomogram should only be used after a single acute acetaminophen ingestion. The line indicates the level at which toxicity is possible after acetaminophen overdose. The level should be plotted in relationship to the time of ingestion to determine the likelihood of toxicity and the need for treatment. Caution should be used in assessing the reliability of the time of ingestion. This nomogram cannot be used for ingestions that occurred greater than 24 hours prior to presentation, repeated supratherapeutic oral ingestions, or iatrogenic intravenous overdose. *Pediatrics* 55(6): 871–876, 197.

SALICYLATES

Acute ingestion of 10-30g can be fatal
Serum levels >2.9 are associated with toxicity

General considerations:

- Must have high index of suspicion; suspect in any patient with tachypnea and metabolic acidosis with no other obvious cause
- Principles of management are **alkalinization of serum and urine** to facilitate renal excretion to prevent movement across the blood-brain barrier, **avoidance of intubation**, and frequent monitoring, keeping in mind potential complications and indications for urgent **dialysis**
- **Indications for dialysis**
  - Altered LOC or seizures
  - Pulmonary edema or cerebral edema
  - Salicylate level >7.2mmol/L in acute ingestion (>3mmol/L in chronic toxicity)
  - Renal failure interfering with excretion
  - Clinical deterioration despite adequate medical therapy
  - Inability to administer bicarbonate, either because of volume overload or severe alkalemia

Clinical Manifestations:

- Vitals: Hyperpnea/↑RR, ↑HR, ↑Temp
- Signs & Symptoms: tinnitus, fever, vertigo, N/V, diarrhea, CNS changes, coma, non-cardiac pulmonary edema
- ABG: mixed respiratory alkalosis and metabolic acidosis; patients may be alkalemic on presentation

Management:

1. **ABC's** – Admit to closely monitored setting and avoid intubation if possible. Consider arterial line
2. **GI decontamination**; activated charcoal 1g/kg up to 50g PO if alert and cooperative, if presenting within 2h of acute ingestion
3. **Alkalinization**: 3 amps of $NaHCO_3$ (1 amp = 50mL of 8.4% sodium bicarbonate) in 1L D5W at 250 ml/hr to target serum pH 7.50-7.59, checking ABGs Q2H (and if possible target urine pH 7.50-8.00)
   - Alkalemia on presentation is NOT a contraindication to bicarbonate infusion; only if severely (pH >7.60) alkalemic
4. **Monitoring**; frequent vitals, Q2H ABGs for pH monitoring, Q2H salicylate levels, frequent measurements of K, Cr, and urine output
   - Monitor for increasing salicylate levels, rising pCO_2 or falling pH indicating respiratory failure, severe alkalemia (>7.60) or hypokalemia with bicarbonate infusion, and rising Cr or oliguria indicating renal insufficiency
   - Correct hypokalemia and hypomagnesemia as it may prevent proper urine alkalinization
5. **Dialysis**; monitor for acute indications; early referral to nephrology will facilitate rapid initiation of dialysis if required

TRICYCLIC ANTIDEPRESSANTS (TCA)

General Considerations:

- Potentially deadly intoxication requiring high index of suspicion- nonspecific signs/symptoms
- Major complications include ventricular arrhythmias and seizures
- Mainstays of therapy are **sodium bicarbonate** for cardiac toxicity and **benzodiazepines** for seizure
- Newer treatment consideration – lipid emulsion

Clinical Manifestations:

- Anticholinergic toxidrome
  - o Hyperthermia, flushing, mydriasis, decreased bowel movements, urinary retention
- CNS: decreased LOC, delirium, seizures
- CV: conduction delay, arrhythmias, sinus tachycardia, hypotension

Investigations:

- History is paramount as levels cannot be obtained immediately
- Urine toxicology screen not useful in acute management
- lytes, HCO3, ABG
- 12- lead ECG (note: significant toxicity can occur even with a normal ECG)
  - o QRS > 100 msec = increased seizures, >160msec = increased VT/VF
  - o Ratio R to S wave in AVR > 0.7 or R wave in AVR > 3mm
  - o deep, slurred S wave in I & AVL

Management:

1. **ABCs**: Admit to closely monitored setting with reliable IV access, early airway protection
2. **GI decontamination**: activated charcoal 1g/kg up to 50g PO if alert and cooperative, if presenting within 2h of ingestion
3. **Sodium Bicarbonate (NaHCO3):**
   - Indications: wide QRS >100msec, ventricular arrhythmia, hypotension
   - Bolus: 2-3 amps (1 amp = 50mL 8.4% sodium bicarb) IV push (watch for resolution of ECG changes), repeat 5 min later if no response
   - Infusion: after IV bolus given, 3 amps in 1L D5W at 250 ml/hr
   - Treatment target: narrow QRS, improving BP, maintain serum pH 7.50-7.55 (check lytes, ABG frequently)
   - Taper bicarbonate infusion once wide QRS resolves
4. **Antiarrhythmics:**
   - For ventricular arrhythmias refractory to sodium bicarbonate
   - Lidocaine or Magnesium (1-2g IV) if unresponsive to NaHCO3; may use hypertonic saline
   - Class I-A, I-C, III agents are contraindicated
5. **Anti-Seizure:**
   - Phenytoin and flumazenil are contraindicated
   - Benzodiazepines (lorazepam 2mg IV x1 repeated q3-5min until seizure resolves) are mainstay of treatment
6. **Rescue antidote**: Lipid Emulsion Therapy (Intralipid)

METHANOL & ETHYLENE GLYCOL TOXICITY

Common sources include home-made liquor, antifreeze, windshield washer fluid, de-icer fluid, paint thinner, varnish
Methanol 40% is lethal at 30-250ml and Ethylene glycol is lethal at 100ml

General Considerations:
- "Toxic alcohols" difficult to clinically distinguish from ethanol ingestion acutely
  - However if not identified and treated can result in severe end-organ toxicity
- Toxicity occurs secondary to metabolites (eg, formaldehyde, formic acid, glycolic acid, oxalic acid) formed via alcohol dehydrogenase
  - Methanol; metabolized to formate, toxic to retina and CNS
  - Ethylene glycol; metabolites cause oxalate crystal deposition and renal failure
- Diagnosis hinges on presence of **anion gap metabolic acidosis** and presence of **osmolar gap**

Clinical Manifestations:
- Methanol
  - Early: drowsiness, nausea, headache, vertigo, mydriasis
  - Late: blurred vision, decreased light reflex, retinal sheen (due to retinal edema), hyperemia of optic disks, relative afferent papillary defect is ominous.
- Ethylene glycol
  - <12hrs: vomiting, ataxia, stupor, and coma with absent reflexes and anisocoria, hypertension, Kussmaul's breathing
  - 12 - 24 hrs – hypotension, myocarditis, pulmonary edema, seizures
  - 24 - 72 hrs – flank pain, renal failure, crystalluria (due to calcium oxalate crystals)

Diagnosis:
- High index of suspicion and good collateral history are important along with the following tests:
- Anion gap metabolic acidosis
- Elevated osmolar gap (measured – [2Na + BUN + Glu])
  - Interpret with caution; late presentation may have normal osmolar gap due to conversion of parent alcohol into smaller metabolites
  - Normal osmolar gap does not rule out late toxic alcohol ingestion
- Serum methanol and ethylene glycol levels
  - Will need medical biochemistry approval for this
- Urine for oxalate crystals; nonspecific finding in late ethylene glycol poisoning

Management:
1. **ABCs**: admit to monitored setting with appropriate IV access
2. **GI decontamination**: unlikely to be of benefit because of rapid absorption and large dose needed
3. **Fomepizole**
   - Mechanism of action: inhibits EtOH dehydrogenase
   - Indications: severe acidosis, end-organ damage, methanol >6mmol/L, ethylene glycol >3mmol/L

- Dose: 15mg/kg in 100ml of D5W over 30 min then 10mg/kg Q12H x 48hrs then 15mg/kg
- Q12H until methanol <6mmol/L and/or ethylene glycol <3mmol/L and pH is normalized
- Ethanol can be used if fomepizole not available, but considered inferior

4. **Hemodialysis**
   - Early nephrology consult if suspicion of toxic alcohol ingestion is high
   - Decision to initiate dialysis governed by several factors:
     - Suspected or known toxic alcohol ingestion
     - Presence of anion gap acidosis
     - Drug level
     - Presence of end-organ dysfunction (ie. Visual changes, renal impairment)
     - Time from ingestion to presentation
5. **Cofactor therapy**
   - Folic acid 1mg/kg IV Q4H x 6 doses (max 50mg per dose) for methanol poisoning
   - Can also give thiamine 100mg IV q6h and pyridoxine 50mg IV q6h for ethylene glycol poisoning

PRE-OPERATIVE ASSESSMENT

NON-CARDIAC SURGERY EVALUATION OVERVIEW

Bedside Assessment:

1. PMHx specifics: chronic diseases (review of organ systems), bleeding history, previous surgeries and complications
2. Medications: last date for anticoagulants/antiplatelets, OTCs, herbal medications, allergies, non-compliance
3. Social history: EtOH (withdrawal risk), smoking, illicit drugs
4. History of presenting illness: Focus on potential medical red flags of presentation (e.g. syncope → fall)
5. Baseline functional capacity - 1 MET = ADL, 4 MET = 1 flight of stairs/light house work, 10 MET = sports (not recommended by the CCS – evidence lacking)
6. Review of Systems:
   - Cardiac: MI/HF hx/symptoms, syncope, hx severe AS
   - Respiratory: smoking pack-years, cessations, disease control, active exacerbations, history of OSA, CPAP compliance
   - Endocrinology: Diabetes type, insulin regimen & compliance, steroid current or recent (1 year) use
   - Hematology: previous thrombosis, antiplatelets/anticoagulants, bleeding history, stroke risk (CHADS)
7. Screening: OSA STOP-Bang score
8. Screening for openness to smoking cessation (recommended by new CCS guidelines)
9. Physical Examination
   - Cardiac: Rule out mod-sev AS, volume status (heart failure, hypovolemia)
   - Respiratory: pneumonia, uncontrolled or exacerbation of underlying asthma/COPD, OSA (neck circumference)
   - Neurologic: screen for delirium, undiagnosed deficits (e.g. strokes)

Cardiac Risk Assessment:

1. Perioperative cardiovascular risk: Provide peri-op MACE – RCRI score (use revised risk estimates as per CCS guidelines), consider general procedural risk, BNP/nt-pro BNP (to be available soon in Hamilton)
2. Aortic stenosis: preoperative ECHO if highly suspected on exam, or no ECHO within 1y for known significant disease (should not delay surgery if urgent or emergent surgery)
3. MINS: Consider post-operative troponin/ECG monitoring

Disease Management:

1. I.E. Prophylaxis: screen for both presence of indication and appropriate procedure
2. Ongoing infections: Treat any ongoing infections (e.g. pneumonia, cellulitis)
3. Heart failure: IF decompensated, treat and consider delay in surgery
4. Aortic stenosis or other severe valvular disease: May require higher monitoring environment, consider cardiology and anesthesia consults
5. UA/NSTEMI, CCS III-IV: consider cardiology consult – may require further investigation if elective surgery. If urgent/emergent surgery, will need to weigh risks and benefits of delaying surgery on a case by case basis.
6. OSA: Screen using STOP-Bang. If moderate to severe by suspicion or history (non-compliant), needs increased post-op monitoring. If elective OR, consider sleep study before surgery
7. Respiratory exacerbation: optimize treatment (e.g. COPD exacerbation meds) and delay surgery if possible
8. Delirium: risk calculators helpful in predicting post-operative course

Medication Management:

1. Antiplatelet agents: if recent ACS or DAPT for PCI, consider cardiology consultation, usually ASA continued if PCI within the last year (as per CCS guidelines). Otherwise hold 5-7 days prior to surgery.
2. Anticoagulants: i.e. warfarin, DOACs. Warfarin requires reversal. DOACs require 2-3d to clear – speak to surgical team about OR date/time and comfort of performing procedure on current medications.
3. Inhalers: continue and/or optimize to prevent exacerbations
4. Steroids: stress-dose if physiologically suppressed (usually prednisone >=5mg for >3 weeks)
5. Insulin-dependent DM: give ½ to 2/3 long acting, hold short acting & oral medications; D5W if NPO ; insulin infusion if long procedure
6. Medications with withdrawal potential: continue opioids, benzodiazepines

Investigations/ "Recommendations" Summary:

1. ECG: most patients
2. CXR: suspected or known pulmonary disease
3. ECHO: suspected AS (not for LV dysfunction)
4. Noninvasive cardiac testing: do not order for asymptomatic patients (no longer recommend METS-based risk stratification)
5. Medication recommendations: give explicit instructions for last dose times/dates
6. Post-op CPAP: if stable, pt to use own. If noncompliant or high suspicion (STOPBang), needs CPAP with increased monitoring (stepdown, CCU, ICU)
7. Post-op cardiac monitoring: troponins and ECG daily x48-72h (on all patients age ≥ 65, or patients age ≥ 45 with previous hx of CAD, PAD or stroke, or patients with RCRI ≥1, or elevated BNP/nt-pro BNP)
8. Post-op hemodynamics monitoring: Consider CCU, ICU, or stepdown monitoring for high peri-operative MACE
9. Post-op telemetry: consider with known arrhythmias. Predictive of ischemia, but resource-intensive. Generally, troponin screening is adequate.

Cardiac Risk by Procedure Type:

Cardiac Risk	Surgery Type
High (>5%)	Emergent, Vascular, Aortic, Large blood loss or fluid shifts
Intermediate (1-5%)	Orthopedic, Carotid Endarterectomy, ENT, Thoracic, Abdominal, Prostate
Low (<1%)	Ambulatory, Endoscopic, Cataracts, Breast, Local anesthesia

Vascular Disease: Aortic Stenosis

Importance:

- Doubles mortality risk (2.1% vs. 1.0%), triples MI risk (3.0% vs. 1.1%). Highest risk if symptomatic (OR 7), symptomatic (OR 3), coexisting severe MR (OR 10)Late: blurred vision, decreased light reflex, retinal sheen (due to retinal edema), hyperemia of optic disks, relative afferent papillary defect is ominous.

Clinical Rational Exam for AS (JAMA):

- If no radiation to the right clavicle, unlikely AS (LR- 0.1).
- If radiation is present, look for presence of below combined findings (0-2: LR+ 1.8; 3-4 LR+40)
  - (1) reduced/absent S2 (2) reduced carotid volume ("parvus") (3) slowed increase of pulse ("tardus") (4) loudest in right 2nd intercostal space
- Other helpful signs to rule in AS: Brachio-radial delay, apico-carotid delay, late peaking murmur, right carotid radiation, effort-associated syncope

Clinical Rational Exam for Severe AS (McGee 4e):

Rule in Severe AS	LR
Delayed carotid upstroke	3.5
Sustained apical pulse	4
Absent/diminished S2	4
Late peaking murmur	4
Prolonged murmur	3

Rule out Severe AS	LR
Brachioradial delay	0.05
Apical-carotid delay	0.05
No radiation to the neck	0.1
Normal peaking murmur	0.2
Lack of prolonged murmur duration	0.2
Blowing Quality	0.1

Management:

- Consider expedited pre-operative ECHO if:
  - Moderate/severe AS previously known with no recent ECHO (within 1y) or worrisome clinical features (angina, syncope, dyspnea)
  - Moderate/severe AS suspected on exam
- Consider delaying or cancelling non-urgent OR and consult cardiology if AS severe (mean gradient > 40mmHg, valve area <1cm²)
  - Surgical or transcatheter AVR may be considered for elective cases but will need to weigh risks and benefits of delay to surgery on QOL and requirement of possible antiplatelet agents post valve replacement.
- For urgent/emergent surgeries, may consider increased intra and post-operative hemodynamic monitoring (cardiology and anesthesia consult)

Non-AS Valvular Disease:

- Severe left sided valvular disease (AR, AS, MR, MS) can be associated with increased perioperative morbidity/mortality
- Determine severity and whether patient is symptomatic
- Consider pre-operative echocardiogram based on clinical examination and if recent imaging is present
- If elective, consider cardiology referral for assessment for repair/replacement prior to OR
- Consider increased intraoperative and postoperative hemodynamic monitoring (cardiology and anesthesia consult)

Preoperative Cardiac Risk Assessment:

Summary of Risk Assessment Strategies:

1. Risk indices: Calculate the RCRI (see below), which is externally validated predictive index based on six clinical risk factors
2. Biomarkers: Consider BNP or NT-proBNP (soon to be available) on all patients age ≥ 65, or patients age ≥ 45 with previous hx of CAD, PAD or stroke, or patients with RCRI ≥1
3. Imaging: Resting ECHO (for LVEF), CCTA (for CAD), exercise stress testing, and pharmalogic stress testing (e.g. MPI) are not recommended as they either do not significantly improve risk prediction and in some cases may even overestimate risk.

Revised Cardiac Risk Index ("Lee" Index)

Predicts perioperative risk of major adverse cardiac events (MACE): MI, pulmonary edema, VF/cardiac arrest, complete heart block

RCRI Predictor	Criteria
High-risk surgery	Intraperitoneal, intrathoracic, suprainguinal vascular
Ischemic heart disease	MI, stress test, angina, nitrate use, Qwaves on ECG
Heart failure	Hx CHF, PND, exam, CXR, S3
Cerebrovascular Disease	Stroke or TIA
Diabetes	Insulin dependent only
CKD	Cr > 177

RCRI Score	Perioperative MACE
0	4%
1	6%
2	10%
>=3	15%

BNP and NT-proBNP

Measure levels if: >65y, RCRI >=1, 45-64y with CVD

Test Result	30d Postop Death or MI
NT-proBNP < 300ng/L BNP < 92 mg/L	5%
NT-proBNP > 300ng/L BNP >92 mg/L	22%
Test Result	30d Postop Death or MI

Preoperative Cardiac Risk Modification:

Medications for Prevention of Perioperative Cardiac Events:

1. ASA: Do not initiate ASA. Increases bleeding without reduction in events. (POISE-2, PEP) Hold for 5-7 days prior to surgery if patient already on ASA (*See Post PCI section if patient has stent*).
2. B-blocker: Do not initiate B-blocker within 24 hours before non-cardiac surgery, but continue in chronic users. Reduces MI but increases death, stroke, hypotension, bradycardia (POISE)
3. A2-agonist: Do not initiate for perioperative prevention of cardiac events. No reduction in MI/death but increases cardiac arrest, hypotension, and bradycardia (POISE-2).
4. Calcium channel blockers: Do not initiate CCBs for prevention of events (poor evidence).
5. ACEI/ARB: Withhold 24 hours prior to surgery in chronic users. Increases intraoperative hypotension without known benefit.
6. Statins: Continue in chronic statin users. Insufficient data at present time to comment on whether a new statin should be initiated in a statin naïve patient.

Coronary Vascularization before Non-Cardiac Surgery:

- Do not prophylactically revascularize patients undergoing noncardiac surgery (CARP – no difference in mortality)
- Consider preoperative revascularization in CCS III-IV or unstable angina, though this must be weighed against perioperative interruption of DAPT with a new coronary stent

Smoking Cessation:

- For elective procedures, discuss and facilitate smoking cessation 8 or more weeks prior to surgery (as per CCS recommendations)

Preoperative Cardiac Event Monitoring:

Overview:

- MINS: Myocardial Injury after Non-Cardiac Surgery – includes MIs and troponin elevation without CP/ECG changes; 95% non-ACS
- Presentation: 93% asymptomatic, often no ECG changes. 94% occur within 48-72h post-OR.
- Prognosis: Perioperative troponin elevation predicts 30-d (VISION study) and 1-y mortality rates, regardless of symptom presence

Monitoring:

- CCS Recommends monitoring for "High Risk" patients = **RCRI >=1, all patients >65y**, elevated BNP/NT pro-BNP, 45-65y with CVD
- Postoperative daily troponin x48-72h on all high-risk (>5% MACE) patients
- Postoperative daily ECG x48-72 on all high-risk (>5% MACE) patients
- Postoperative telemetry – may be unnecessary in light of hs-Troponins. May consider if patient high risk for unstable arrhythmias.

Management of MINS:

- Lifelong daily ASA 81mg po daily
- Lifelong daily statin (e.g. atorvastatin 80mg po daily)
- Recommendations based on reduced 30-day mortality in prospective cohort studies

Overview of Perioperative Cardiac Guideline Updates:

- This section was based on a combination of recommendations by the ACC/AHA (2014) and CCS (2016). The CCS guidelines provide recommendations based on the most current evidence base, and should be used where discrepencies arise with previous guidelines.
- In the latest CCS update, RCRI MACE risk estimates are increased compared to previously reported estimates. This is based on the application of the RCRI to data sets where non-elective procedures were also included, and where biomarker screening with troponin was routinely used to screen for events.
- In the latest CCS update, BNP/NT-pro BNP was recommended as an adjunct to perioperative risk stratification – this test is likely to be available in the near future at HHS/SJHH. This test is a new recommendation compared to previous guidelines.
- Contrary to the ACC/AHA, there is no recommendation by CCS guidelines on use of METS to predict perioperative events. There is a lack of evidence to support the use of self-reported functional capacity, with current studies suggesting self-reported functional capacity are not predictive of major perioperative cardiac complications.
- CCS recommends the use of ASA and statins for MINS. This is based POISE substudy data and not mentioned in previous guidelines.

Cardiology Post-PCI Antiplatelet Therapy:

Emergent/Urgernt Surgery:

- ASA: should be continued if possible if stent in the last year.
- P2Y12 platelet receptor inhibitor: whether continued depends on risks/benefits discussed with involved teams. Consider cardiology consult.

Elective Surgery:

- DES: Delay surgery for at least 3 months (and ideally 6 months) after DES implantation
- BMS: Delay surgery for at least 30 days after BMS implantation

Infectious Disease:
Endocarditis Prophylaxis

Indications	Procedures	Antibiotics
• Prosthetic cardiac valve/repair • Previous infective endocarditis • Cardiac transplant with valve regurgitation from abnormal valve • Congenital heart disease 1) unrepaired cyanotic, including palliative shunts/conduits) 2) repair with prosthetics, within 6mo sx 3) repair with residual defects at or adjacent to prosthestic material	• Dental: gingival or periapical manipulation, oral mucosa perforation • Respiratory: incision or biopsy of respiratory tract mucosa (tonsillectomy, adenoidectomy, bronchoscopy w biopsy) • Not recommended for GI/GU procedures	• Take all 30-60 min pre-procedure • First line: Amoxicillin 2g PO x1 • Allergy: Cephalexin 2g PO x1 • Allergy: Clindamycin 600mg PO x1 • Cannot PO: Ampicillin 2g IM/IV • Cannot PO, has allergy: cefazolin 1g IM/IV, ceftriaxone 1g IM/IV, clindamycin 600mg IM/IV

Respirology:

- Known OSA:
  - Compliant, stable: bring machine in for use. Does not need additional monitoring
  - Non-compliant: May require additional monitoring depending on severity
- Undiagnosed OSA Screen: If high risk moderate-severe OSA, consider post-operative CPAP and increased monitoring (ICU)
  - High-risk Criteria:
    - Score 5+
    - Score of 4, including 1 of: male, BMI > 35, neck size enlargement
  - **STOPBANG** Score (1 point each):
  - Snoring, Tired, Observed, Pressure (Hypertension), BMI > 35, Age > 50, Neck size at Adams apple (Male: 43cm, female, 41cm), Gender (male)
- Smoker: Attempt cessation, best 8 weeks prior to surgery
- COPD/asthma: optimize control with inhaled bronchodilators and glucocorticoids, administer systemic steroids if flare
- Pulmonary infection: provide antibiotic treatment and delay elective surgery
- General respiratory manoeuvres: aerobic exercise, breathing exercise, inspiratory muscle training
- Medications: continue inhalers on day of surgery

Hematology:

See Hematology section on perioperative anticoagulation management and anticoagulation reversal

Endocrinology:

Diabetes:

- Goals: Avoid hypoglycemia, prevent DKA/HHS, maintain fluid/lytes balance
- Glycemic targets: 4.5-10 (ADA) is reasonable (conventional better than tight control)
- T2D, diet controlled: sliding scale Q6H
- T2D, medications: hold AM of surgery, sliding scale Q6H
- Resumption of medications: resume insulin & PO meds once eating well

Insulin Dependent DM (T1 or T2):

- Omit rapid acting insulin if applicable
- Start D5W @ 75-125cc/h once NPO
- Check CBG Q6H, adjust level with sliding scale
- Give ½-1/3 of AM intermediate/long-acting insulin dose
- If on infusion pump, continue the basal rate
- For complex procedures, intraoperative IV insulin is needed

Insulin Infusions:

- Indications: emergency surgery with hyper/hypoglycemia, lengthy procedures (CABG, renal transplant, neurosurgical)
- Starting rate: Blood glucose / 5 (round to nearest 0.5)
- Adjustment: Based on CBG Q1H & available protocols
- NEVER stop insulin infusions in T1D (ketosis risk)
- Post-operative cessation: once pt eating, stop 2h after receiving subcutaneous dose

Steroids:

1. Determine if patient requires stress dosing due to chronic steroid use (>3wk) at any time in the last 12 months:
   - If prednisone <5mg (or equivalent) po daily, not suppressed
   - If prednisone >20mg po daily for > 3 weeks, suppressed
   - If prednisone 5-20mg po daily for >3 weeks, indeterminate
2. Determine if further testing is required:
   - If not suppressed, no stress dosing or testing required
   - If indeterminate and elective surgery, screen for adrenal insufficiency → AM cortisol +- ACTH stimulation test
   - If indeterminate but urgent surgery, proceed to stress dosing
   - If suppressed, proceed to stress dosing
3. Determine stress doses required (based on surgical stress):

Surgical Stress	Examples	Morning Dose	IV Hydrocortisone On call to OR	Post-Procedure IV hydrocortisone
Mild	Hernia Repair	Give	None	None
Moderate	Vascular (extremity), THA, TKA	Give	50mg	25mg q8h x24h
Severe	Gastrectomy, Cardiac surgery	Give	100mg	50mg IV q8h x24h

Delirium:

Risk	Points
Age >70	1
EtOH Abuse	1
Cognitive Impairment	1
Severe physical impairment	1
Abnormal pre-op labs	1
AAA surgery	2
Non cardiac thoracic surgery	1

Score	Delirium Risk	
0	1-2%	Low
1-2	8-19%	Moderate
3	45-50%	High

- Prevention: early Foley d/c, control pain (multimodal, minimize narcotics), bladder, hypoxia, less benzo's, re-orientation
- EtOH Hx: Thiamine and CIWA protocol and monitor for withdrawal

MEDICATIONS MANAGEMENT

General Principles:

- Considerations may be different between elective vs. urgent surgical procedures
- Consider discontinuation of most non-essential medications
- Evidence for continuation vs. discontinuation is weak for many drugs – individualize based on patient-specific risks vs. benefits
- Speak directly with surgical team in regards to planned OR time and whether procedure can be done on antiplatelet/anticoagulant medications
- Avoid stopping medications with high risk of rebound/withdrawal symptoms
- Do not miss substance use (alcohol, nicotine, illicit drug use) on history
- The below list refers to continuing or holding medications taken chronically, not whether to start a new medication (see cardiac risk assessment – medications section above)
- When both "Give" and "Hold" columns are checked, the choice will depend on each individual patient
- Perioperative hypotension is emerging as a significant cause of downstream morbidity and mortality. Thus in general, anti-HTN medications should be held 24 hours prior to surgery and not restarted until 48-72 hours postoperatively when it is clear that patient is no longer hypotensive.

Cardiovascular	Give	Hold	Comments
Beta Blockers	✓		Avoid hypertension, tachycardia, MI
Alpha-2 Agonists		✓	Avoid hypotension
CCBs		✓	?Reduced arrhythmia, MI (poor data)
ACE-I/ARBs		✓	↑ Intraop hypotension vs. worsen HTN or HF (poor data)
Diuretics		✓	Hypokalemia, hypotension
Statins	✓		Lowers mortality (observational)
Non-statin lipid drugs		✓	Risk rhabdomyolysis (poor data)
Digoxin	✓		May reduce post-op SVT (poor data)

Respiratory	Give	Hold	Comments
Inhalers (All Types)	✓		Reduces post-op pulmonary complications
Theophylline		✓	Risk arrhythmia, neurotoxicity when supratherapeutic periop
Systemic steroid	✓		Risk of adrenal insufficiency
Leukotriene Inhibitors	✓		No known harm or withdrawal symptoms

Endocrine	Give	Hold	Comments
Systemic steroid	✓		Risk of adrenal insufficiency; stress dose
Metformin		✓	Risk of renal hypoperfusion, tissue hypoxia (lactic acidosis)
Sulfonylureas		✓	Risk of hypoglycemia
Thiazolidinediones (TZD)		✓	Risk of heart failure
SGLT2-Inhibitors		✓	Risk of hypovolemia
DPP-4/GLP-1		✓	Risk of altered GI motility
Regular rapid-acting Insulin		✓	Use sliding scale as pt is NPO
Long-acting insulin	✓*		*Give modified last dose (1/2-2/3)
Insulin Pump	✓		Continue current basal rate
Levothyroxine	✓		Parenteral if total NPO >5d
Methimazole/Parathyrouracil	✓		Maintain euthyroidism
Bisphosphonates	✓		Must hold >3mo for effect – ?hold for elective dental surgery (jaw necrosis)
Oral contraceptives		✓	Thrombosis risk; hold 4-6wk prior; B-hCG
Hormone replacement therapy	✓*	✓*	Thrombosis risk; *hold 4wk if mod-high risk of thromboembolism
SERMs for Malignancy	N/A	N/A	In consultation with oncology

Hematology	Give	Hold	Comments
ASA (part of DAPT for post PCI†)	✓		Continue if recent within 1 year
ASA		✓	Hold 7d prior
P2Y12 Platelet Inhibitor, DAPT for PCI†	✓	✓	Discuss with cardiology – surgical bleed risk vs. IST/MI/death
P2Y12 Platelet Inhibitor		✓	Hold clopidogrel, ticagrelor 5d.
Aggrenox (ASA/dipyridamole)		✓	Hold 7-10d prior – minimal data
Cilostazol		✓	Hold 2-5d prior to elective surgery
NSAIDs		✓	Hold 3d prior (1d for ibuprofen); bleeding, renal dysfunction
Warfarin		✓	Hold 5d prior – reverse if needed
DOACs		✓	Hold 2-3d prior

† Consider cardiology consult post-PCI if have not received minimal DAPT time for stent (BMS 30d, DES 3-6mo)

Psychiatry	Give	Hold	Comments
TCAs	✓		Depression/pain risk vs. peri-op arrhythmias; taper 7-14d if sig. cardiac disease
SSRIs, SNRIs, buproprion	✓		Small risk of perioperative bleeding
MAO Inhibitors		✓	May be needed for control of severe psychiatric disease. Anesthesia must use MAO-safe procedures; risk of hypertensive crisis, serotonin syndrome, opiate overdose intraop
Lithium, Valproate	✓		Li – fluid/lytes if nephrogenic DI
Antipsychotics	✓	✓	Depends on psychosis risk. Watch QTc, rebound psychosis, withdrawal
Benzodiazepines	✓		Avoid rebound if chronic user
Psycho-Stimulants		✓	Hypertension, arrhythmia, seizure threshold, medication interactions

Other Meds	Give	Hold	Comments
PPIs/H2RAs	✓		Reduced risk of gastric aspiration chemical pneumonitis
Chronic Opioid Use	✓		Avoid rebound; consider increased dose or short-acting or transdermal
Methadone	✓		If cannot take PO, give ½-dose IV; give breakthrough opioids (high dose-resistance)
Anti-Epileptics	✓		Avoid intraoperative seizures
Carbi/Levodopa	✓		PD flare reduction; short ½-life
Dopamine agonists	✓		PD flare reduction
Pyridostigmine		✓	Avoid muscarinic side effects
Methotrexate	✓	✓*	*Hold 2wk if: renal impairment,

			bone marrow suppression, active infection
Rheumatologic antibodies	✓*	✓*	*Risk of autoimmune flare vs. infection; consider hold 1-2 cycles
Hydroxychloroquine	✓		Minimal risk
Gout drugs (colchicine, allopurinol)		✓	Muscle weakness, polyneuropathy, limited data
Anti-retrovirals	✓		Stopping several days is unlikely high risk in HIV

Herbals	Give	Hold	Comments
Ephedra		✓	Hold 24h; MI risk
Garlic		✓	Hold 24-48h; Bleeding risk
Ginkgo		✓	Hold 24-48h; Bleeding risk
Ginseng		✓	Hold 7d; bleeding risk
Kava		✓	Hold 24h; increased sedative effect
St. John's Wart		✓	Hold 5d; P450 induction
Valerian		✓	Benzo-like withdrawal; taper pre-op or treat with benzo's post-op

RESPIROLOGY

DYSPNEA

Pathophysiology	Etiologies
Airway Obstruction	Asthma, COPD, bronchiectasis, cystic fibrosis, tumor, foreign body
Parenchymal disease	Cardiogenic and non cardiogenic pulmonary edema, ILD, pneumonia, pulmonary hemorrhage, malignancy, ARDS
Vascular	PE, other emboli (tumor, fat), pulmonary hypertension, vasculitis
Pleural disease	Pneumothorax, pleural effusion
Neuromuscular and chest wall disorders	Neuromuscular disease (ALS, muscular dystrophy), obesity, deconditioning, pregnancy, kyphosis, scoliosis
Systemic	Hematologic: severe anemia Metabolic: hyperthyroidism, metabolic acidosis Toxicities: CO poisoning, methemogloblinemia, salicylate poisoning
Physiological	Anxiety, hyperventilation

Evaluation:

1. Assess ABC: Assess vitals, LOC, work of breathing, level of distress (tripoding, accessory muscle use, central cyanosis, pursed lip breathing, nasal flaring)
   - If there is moderate – severe work of breathing **call for help (senior, RT)**
2. History: Acute vs. chronic, positional dependence, with activity or rest, aggravating or alleviating factors, infectious, HF, malignancy symptoms.
3. Physical: Full exam, focus on cardiac, respiratory and volume status.
4. Diagnostic tests are based on the clinic suspicion. Baseline investigations: CBC, Lytes, Cr, VBG, CXR, ECG, further testing is based on the working diagnosis. (I.e. ?HF – echo, ?ILD – HRCT)
5. Treatment: dyspnea is a symptom → need to treat underlying process

HYPOXIA

Hypoxia: low oxygen at tissue level
Hypoxemia: low level of oxygen in the blood

1. Levels correlate with certain exceptions
2. If SaO_2 from pulse oximeter is much higher (>5%) than the expected PaO_2 from blood gas, consider another oxygen-like molecule being detected by the pulse oximeter (carbon monoxide poisoning, methemoglobinemia, or sulfhemoglobinemia)

PaO2 (mmHg)	Expected O2 Saturation (%)
10	10
20	34
30	57
40	74
50	85
60	91
70	94
80	96
90	96.7
100	97.5
>100	>98

A-a Gradient = $FiO2(P_{ATM}-P_{H2O})-P_{CO2}/0.8)$. Assuming P_{ATM} = 760mmHg and P_{H2O}=47mmHg and FiO2=0.21 then → A-a gradient = 150 – PCO2/0.8
An elevated A-a gradient indicates hypoxia is related to pulmonary causes
Normal value < 15 mmHg but increases with age.

Hypoxemia with high A-a gradient:

1. VQ Mismatch
2. Shunt
3. Diffusion Impairment

Hypoxemia with normal A-a gradient:

1. Hypoventilation
2. Reduced partial pressure of O2 (high altitude)

HYPERCAPNIA

Etiology:

1) **Decreased respiratory drive** - "won't breathe": drugs, trauma, increased ICP, CNS infections, brain stem stroke, central apnea
2) **Respiratory failure** - "can't breathe": Neuromuscular disorders (myasthenia gravis, GBS, Poliomyelitis, muscular dystrophies, ALS, myopathies), chest wall disorders, airway (asthma, COPD), parenchyma (ARDS, pulmonary edema, pneumonia, ILD), pneumothorax, ARDS

Evaluation and treatment based on underlying etiology.
Most importantly, need to determine if patient is protecting airway to decide acute management. (NIPPV vs. intubation)
Please see sections on acute respiratory failure in this + ICU section.

APPROACH TO PULMONARY FUNCTION TESTS AND SPIROMETRY

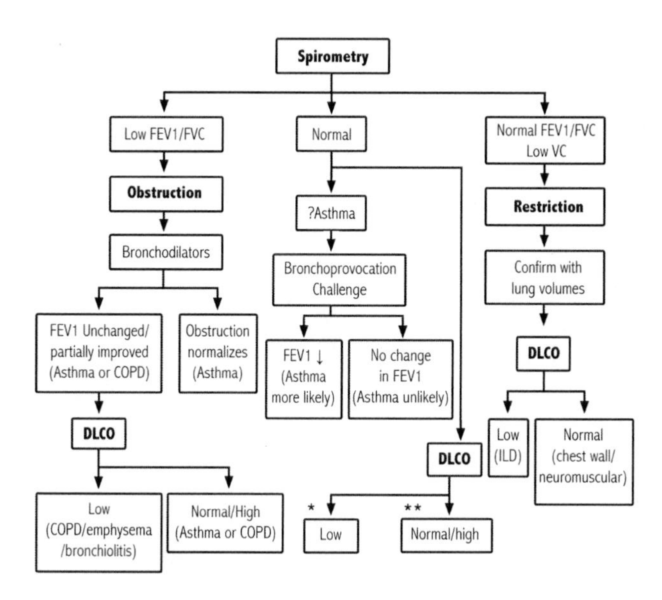

- Normal FEV1 80%-120% of predicted.
- Reduced FEV1/FVC is < 70% (<5th percentile lower limit normal).
- Reduced VC and TLC defined as < 80% predicted.
  * High DLCO and normal lung volumes associated with asthma, obesity and pulmonary hemorrhage
  ** Decreased DLCO and normal lung volumes associated with anemia, PE, pulmonary hypertension, early ILD or early emphysema

ASTHMA EXACERBATION

Definition

Asthma: Characterized by symptoms such as wheezing, shortness of breath, chest tightness and cough in the setting of variable expiratory airflow obstruction, airway hyper-responsiveness and airway inflammation.

Spirometry definition: airflow obstruction with FEV1/FVC <LLN **AND** one of:

1. FEV1 improvement >12% **AND** 200mL post bronchodilator

2. or if normal spirometry,
   a. Positive methacholine challenge test, defined as a decline in FEV1 of >20% upon administration of up to 16 mg/ml of methacholine **OR**
   b. Peak Expiratory Flow variability of >20% on multiple daily readings.

Precipitants
Infection (RSV, adenovirus, atypical bacteria), **allergens** (seasonal, pets, dust mites, molds), **irritants** (smoke, perfumes, emotions, food), **exercise, medications** (ASA, NSAIDs and beta-blocker), **occupational exposures** (allergens or irritants), and **medication non-compliance**

Diagnostic Approach
History: variable
Inflammation/bronchitis: increased cough, sputum
Airway narrowing: wheeze, dyspnea, chest tightness, tachypnea
Risk factors: previous ER visits, ≥ 1 exacerbations in past 12 months, hospitalizations, lifetime previous episode of intubations/ICU, oral steroids, comorbidities (cardiovascular, chronic lung disease), frequent use of "rescue" bronchodilators, poor adherence to prescribed treatment, incorrect inhaler technique, ongoing exposure to triggers

Differential Diagnosis/Asthma Mimickers
Upper airway: laryngospasm/airway stenosis, anaphylaxis, vocal cord dysfunction, tracheal stenosis/malacia
Central airway: tumor, FB aspiration, mucous plug
Peripheral airway: COPD, bronchiectasis, CF, GERD, carcinoid, bronchiolitis/croup in children
Alveolar space: pneumonia, CHF
Vasculature: pulmonary embolus, EGPA (note: should be more specific when talking about vasculitis as it is a broad term. EGPA generally accepted as a mimicker)
Pleura: pneumothorax, pneumomediastinum

History: current symptoms (infectious prodrome, puffer compliance, any potentially allergenic trigger), assess baseline control: compliance, puffer technique, daytime/night time symptoms, physical activity, exacerbations (previous intubation ever)
Exam: Vitals (HR, RR, O_2 requirement), mental status, cyanosis, signs of respiratory distress (pursed-lip breathing, accessory muscle use, paradoxical breathing, diaphoresis), assess air entry, wheezing with prolonged expiratory phase, pulsus paradoxus. **NB: Silent chest and paradoxical breathing indicate severe impairment – consider the need for intubation.**
Investigations:
1) CBC, electrolytes, sputum gram stain and C&S, NPS
2) ABG: normal or high pCO_2 indicates respiratory muscle tiring and potential need for intubation.
3) PEF <40% or <200L/min is severe
4) CXR: look for infiltrates, pneumothorax
5) ECG

Non-ER testing: spirometry, PFTs, methacholine challenge, skin testing, sputum differential cell count (eosinophilic vs. neutrophilic inflammation). These tests are performed after recovery from an exacerbation in the outpatient setting.
Acute Management: Monitored setting including continuous SpO2, IV access, fluids Frequently reassess especially in first hour and consider stepping down therapy as appropriate

	Mild	Moderate	Severe	Near death
History	Cough, dyspnea on exertion, nocturnal awakening, ↑β-agonist with good response	Dyspnea at rest, cough, congestion, chest tightness, nocturnal sx, ↑β-agonist > q4h with partial relief	Laboured, agitated, diaphoretic, difficulty speaking, tachycardia, no relief with β-agonists	Exhausted, confused, ↓LOC, cyanotic silent chest, ↓Resp effort, ↓HR. **GET HELP!!**
Spirometry (predicted) FEV1 PEFR	> 60% > 2.1 > 300	40 - 60% 1.6 - 2.1 200 - 300	< 40% < 1.6 < 200	Unable to perform O2 sat < 90% with O2 suppl.
O2 suppl to keep SaO_2 >92%	Yes/NO	Yes	100%	Intubate (rapid sequence induction) with O_2
Intubation	NO	NO	NO	Yes*
Salbutamol (MDI with A/C)	MDI 4-8 puffs q15min X 3 then R/A	MDI 4-8 puffs q15min X 3 then R/A	Neb 5mg (1mL) in 3mL NS q15min X 3 then R/A; 2.5-10mg q4H PRN	MDI 8 puffs via ventilator X 3 q15min OR Continuous neb at 10 mg/hr
Ipratropium bromide (MDI with A/C)	NO	+/- MDI 4-8 puffs q15-30min X 3, then R/A	MDI or neb 250-500μg (1-2 mL) in 2-3mL NS q15-30min X3 then R/A	MDI 8 puffs via ventilator q15-30min X 3 OR Continuous neb at 1 mg/hr
Steroids*: MDI, Prednisone 40-60 mg po or Methylpredniso-lone 125mg IV	MDI + PO steroids	MDI + PO or IV steroids	MDI + IV steroids	MDI + IV steroids
Magnesium sulfate (requires cardiac monitoring)	NO	NO	YES Magnesium 2g in 250 NS IV over 20min	YES Magnesium 2g in 250 NS IV over 20min
CXR	NO	Yes	Yes	Yes
Admission	Consider d/c home if rapid improvement with 48h FU	Ward	Step down	ICU Consider use of heliox

- IV steroids onset 1hr, peak 5hr, half-life 24hr; PO steroids onset 3h, peak 12hr, half-life 24hr; Inhalers onset 1-2 wks.
- Consider epinephrine if anaphylactic or unable to take inhalers – 0.3-0.5mL of 1:1000 solution IM or SC

ACUTE EXACERBATION OF COPD (AECOPD)

Definition: An acute event characterized by worsening dyspnea, sputum production or sputum purulence that is beyond normal day-to-day variations and leads to a change in medication.

Precipitants: Infections (viral and bacterial) lifestyle/environmental, PE, pollution

Differential Diagnosis: MI, CHF, PE, pneumonia, aspiration, concomitant asthma, pneumothorax

Diagnostic Approach:
- **History**: increased cough, increase or change in colour of sputum, increased dyspnea, tachypnea, constitutional symptoms, precipitants, vaccination history, home O2, number of previous episodes, comorbidities, previous need for non-invasive or invasive ventilation
- **Risk Factors**: FEV1, age, productive cough, length of COPD, recent steroid or antibiotic therapy, # of exacerbations/COPD-related hospitalization in past year, previous intubations, multiple co-morbidities, pulmonary HTN, GERD
- **Physical Exam**: mental status, respiratory distress, tachypnea, paradoxical chest wall/abdo movements, pursed-lip breathing, use of accessory muscles, asterixis, crackles or wheezing on auscultation, volume exam
- **Investigations**: CBC, lytes, ABG/VBG, CXR, ECG, sputum Gram stain + culture, NPS

Acute Management
- Telemetry monitoring and frequently reassess
- **Oxygenation**: Oxygen to keep O2 saturation >92%, or 88-92%, if CO2 retainer. If requiring high FiO2 to correct hypoxemia consider other causes (PE, severe pneumonia, pulmonary edema, ARDS)
- **Short-acting ß-agonist**: Ventolin 2.5 mg by nebulizer q1-4H, or Ventolin 4 puffs by MDI via A/C q4 hours while awake + PRN max q1H
- **Short-acting anticholinergic**: Ipratropium 2 puffs MDI via A/C q4H while awake + PRN max q1H
- **Steroids**:
  - **Prednisone 40 mg PO daily for 5 days.** IV steroids if concern re: GI absorption or patient ventilated.
- **Antibiotics**:
  - Indications – consider whether this is a simple or complicated AECOPD
  - **Simple**: <3 exacerbations/year, FEV1>50% - Organisms include *Haemophilus influenzae*, *Moraxella catarrhalis*, and *Streptococcus pneumonia*. First line antibiotics consider **Azithromycin 500mg PO x 1 then 250mg daily X 4 days.** Alternatives first lines choices: Amox-Clav, Cefuroxime, Doxycycline
  - **Complicated**: Similar symptoms to simple exacerbations plus at least one of FEV1<50%, ≥ 4 exacerbations/year, ischemic heart disease, use of home oxygen, chronic steroid use, antibiotics in the last 3 months. In this case organisms may also include gram negative organisms. Consider treatment with **Levofloxacin 750 mg PO/IV daily x 5 days** (if normal renal function, or dose based on CrCl) or **Moxifloxacin 400 mg PO/IV daily x 7 days**.
  - NB: choice of antibiotics depends on local bacterial resistance patterns.

- Length of treatment: usually 5-10 days, depending on response to treatment.

> Typical indications for antibiotics are the "**Winnipeg criteria**"
> 1. All 3 of increase in dyspnea, sputum volume and sputum purulence
> 2. 2 of above if one of them is sputum purulence (less evidence for this)
> 3. Require mechanical ventilation

- **NIPPV**: Consider BIPAP for patients with the following present: severe dyspnea with clinical signs suggestive of respiratory muscle fatigue, pH ≤7.35, PaCO2 >45mmHg, RR > 25.
  - Call RT to set-up. Perform trial for 2 hours, repeat ABG/VBG in 2 hours to assess efficacy (PaCO2), then PRN. Consult Respirology service if advice required.
- **Invasive Ventilation**: Consider if contraindications to, or failure of NIPPV. Please refer to ICU sections on NIPPV and Invasive ventilation for full details.

PLEURAL EFFUSION

Differential Diagnosis:

1. **Transudative:** (systemic factors – low oncotic pressure, high PCWP): heart failure (80% bilateral), cirrhotic liver disease, hypoalbuminemia, nephrotic syndrome, hypothyroidism, constrictive pericarditis
2. **Exudative:** (local factors – change in pleural surface permeability): parapneumonic, malignancy, pulmonary embolism, rheumatoid arthritis, pancreatitis, post-CABG, chylothorax, esophageal rupture, TB, SLE, aortic dissection, Hemothorax, Meigs' syndrome (benign ovarian tumour), post radiation

History:

- In general: dyspnea, cough, pleuritic chest pain
- Transudative effusions: often are associated with history of renal, cardiac or liver disease, symptoms of volume overload (orthopnea, PND, peripheral edema)
- Malignant effusions: constitutional symptoms (fever, night sweat and wt loss), history of malignancy, hemoptysis
- Parapneumonic effusion/empyema: fever, recent hx of pneumonia
- Hemothorax/Chylothorax: trauma, iatrogenic
- Pulmonary Embolism: DVT symptoms, see PE section for more details on history

Physical Examination:

- Cyanosis, clubbing, tracheal deviation away from a large effusion, asymmetric chest expansion, dullness to percussion, decreased B/S, reduced tactile fremitus
- Examine for extrapulmonary manifestations of differential diagnosis

Investigations:

- CXR: assess size of effusion, ?loculated or free flowing (layers on lateral decubitus)
- Thoracentesis indication: All new pleural effusions (unless evidence of renal/cardiac/liver failure and responsive to Tx) require diagnostic thoracentesis. Large effusions may require therapeutic thoracentesis /rarely chest tube
- Indication for CT chest: exudative effusion of unknown etiology or in case of suspected PE (note effusions are usually <1/3 hemithorax in setting of PE)
- Consider Respirology consultation if etiology of effusion remains unclear or chest tube insertion/management is required
- *See procedure manual for instruction on how to perform thoracentesis.

Pleural Fluid Analysis:

- Send for:
  1) Cell count & differential, gram stain, C&S, AFB (if TB suspected)
  2) LDH protein, glucose, albumin, pH
  3) Cytology – send as much as possible
  4) Flow cytometry – if considering lymphoproliferative cause
- Send for cholesterol, triglycerides if chylothorax suspected, amylase if pancreatitis or esophageal rupture suspected, adenosine deaminase if query TB.
- Simultaneously send serum LDH, total protein albumin.
- NB: use of pleural-serum albumin gradient may help correctly identify exudative effusions beyond Light's criteria (eg. in setting of recent diuresis)

Light's criteria: Exudative if ONE of the following

1) Pleural : Serum Ratio LDH > 0.6
2) Pleural : Serum Ratio Protein >0.6
3) Pleural fluid LDH > 2/3 ULN serum LDH

Characteristic	Examples
Appearance	pus (empyema), blood (?malignancy, trauma), milky (chylothorax), pale yellow (transudate)
Neutrophils	predominate in parapneumonic effusions, pancreatitis, RA, TB, immediate post CABG
Lymphocytes	predominate in malignancy, TB, lymphoma, sarcoid, RA
Pleural glucose < serum	Parapneumonic, empyema, TB, malignancy, RA, esophageal rupture
pH	Normal pH 7.6 Low pH can be seen in empyema, RA, TB, malignancy

Management: *if symptomatic consider therapeutic thoracentesis

1. **Transudative:** O2, diuresis, manage underlying cause, large effusion may require drainage
   a. Cirrhosis – role for chest tube unclear. If fails diuresis, dietary salt restriction → discuss with GI; consider TIPS procedure, candidacy for liver transplant.
2. **Exudative:** (broad differential diagnosis). Most common:
   a. **Malignancy** – various therapeutic options including repeat thoracentesis, pleurodesis or tunneled pleural catheter insertion
   b. **Parapneumonic/empyema** – presence of large effusion (>1/2 hemithorax), positive gram stain or culture or pH < 7.2 are indications for chest tube insertion. All require antibiotics. Some patients may benefit from intrapleural administration of t-PA and DNAse
   c. **TB :Treatment duration is 6 months with standard anti-TB regimen. Consult Infectious Diseases**
   d. **Rheumatoid effusion** – may mimic parapneumonic effusion on biochemical analysis.

ARTERIAL BLOOD GAS INTERPRETATION

Overall Approach to Acid-Base:

1. Normal values:
   pH: 7.40 (7.35 – 7.45)
   pCO_2: 40 mmHg (35 – 40 mmHg)
   [HCO_3]: 24 mmol/L (21 – 28 mmol/L)
   Anion Gap (AG) = Na – (Cl + HCO_3) = 12 ± 2 mEq/L

2. Verify data: [H+] = pCO2 X 24/[HCO3]

pH:	6.90	7.00	7.10	7.20	7.30	7.40	7.50	7.60
[H+]:	125	100	80	64	50	40	33	25

x0.8 (100 → 80); x1.25 (25 → 33)

3. Basic diagnostic algorithm

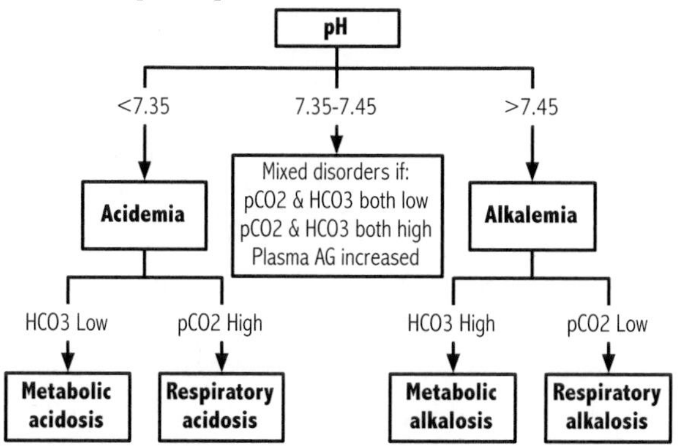

4. Respiratory and renal responses to primary acid-base disorders

Primary Disorder **(bold)**	pCO2	HCO3	ΔpH
Acute Respiratory Acidosis	↑**10**	↑1	↓0.08
Chronic Respiratory Acidosis	↑**10**	↑3	↓0.05
Acute Respiratory Alkalosis	↓**10**	↓2	↑0.09
Chronic Respiratory Alkalosis	↓**10**	↓5	↑0.02
Metabolic Acidosis	↓10	↓**10**	↓0.08
Metabolic Alkalosis	↑6	↑**10**	↑0.09

5. Assess for additional metabolic abnormalities – calculated delta AG and add it to HCO3, if <23 a non AG metabolic acidosis is also present, if >30 a primary metabolic alkalosis is present, or if 23-30 no other metabolic abnormalities present

METABOLIC ACIDOSIS

1. **Look for respiratory response:** ↓1 mmHg $PaCO_2$ for every 1 mmol/L ↓HCO_3. Suspect mixed disorder if significant deviation from this ratio.

2. **Look for anion gap (AG)**
   AG = Na – (Cl + HCO_3)
   Normal 8-12. Adjust for albumin → for every 10 ↓albumin, AG ↓2.5

   Increased AG metabolic acidosis – DDx: MUDPILES CAT

   Methanol, **M**uscle (rhabomyolysis)
   Uremia
   Diabetes and other ketotic states (EtOH, starvation)
   Paraldehyde
   Isopropyl alcohol, **I**NH, **I**ron
   Lactate
   Ethylene glycol, **E**thanol
   Salicylates, **S**trychnine, **S**epsis
   Cyanide
   Arsenic
   Toluene

   Osmolal gap = measure serum osmolality – (2 X [Na] + [urea] + [glucose]); normal ≤ 10; clinical important causes of increased gap: methanol, ethanol, diuretics (mannitol, sobitol, and glycerol), isopropanol, and ethylene glycol. Methanol and Ethylene glycol will cause both metabolic acidosis and increased osmolar gap. Others not associated with metabolic acidosis (ethanol acidosis usually due to lactic acid).

   NOTE: normal gap does *not* exclude toxic levels of methanol or ethylene glycol

   Normal AG metabolic acidosis – DDx:
   $U_{AG} = (U_{Na} + U_K - U_{Cl})$

   Negative U_{AG} (increased renal NH_4^+ excretion → *appropriate* renal response):
   - **GI bicarb loss**: obstruction, diarrhea, billiary drainage, fistula
   - **Urinary tract diversion**: uterosigmoidostomy, ileal conduit
   - **Exogenous**: HCl, NH4Cl, CaCl2, hippuric acid (glue sniffing), excessive NS, TPN, cholestyramine
   - Post hypercapnia

   Positive U_{AG} (failure of kidneys to excrete NH_4^+):
   - **Failure to make new bicarb**: inability to excrete NH4Cl
     - Distal (type I) RTA: impaired H+ secretion, urine pH > 5.3; Sjogren's, multiple myeloma, lupus, amphotericin, hepatitis
     - Aldosterone deficiency/resistance (type IV) RTA: impaired H^+ and K^+ secretion and ammonia production, urine pH < 5.3, serum K elevated;

often caused by diabetes, NSAIDs, ACE-I, trimethoprim, heparin, cyclosporine, Addison's, interstitial renal disease, lupus nephritis
    - Early renal failure
- **Renal bicarb loss**: Proximal tube RTA (type II): Diamox, Fanconi, amyloidosis
- Think about RTAs in outpatients with normal anion gap acidosis unexplained. In patients who comes unwell to the hospital with normal anion gap acidsis, consider other causes.

3. Management: Rx underlying causes

When to use bicarbonate drip (2-3 amp NaHCO3 per litre D5W):

- Lactate acidosis: give 1-2 mM/kg if severe ($pH \leq 7.15$); treat underlying cause
- DKA: no role unless severe cardiac suppression ($pH < 7.10$).
- EtOH: only when severe acidosis unresponsive to fluid resuscitation.
- Methanol/Ethylene glycol: no role for HCO_3; in life-threatening acidosis, initiate hemodialysis.
- RTA's: HCO3 and K for RTA1
- Dilution or loss: HCO3 suppl.

METABOLIC ALKALOSIS

1. Etiology and Diagnosis

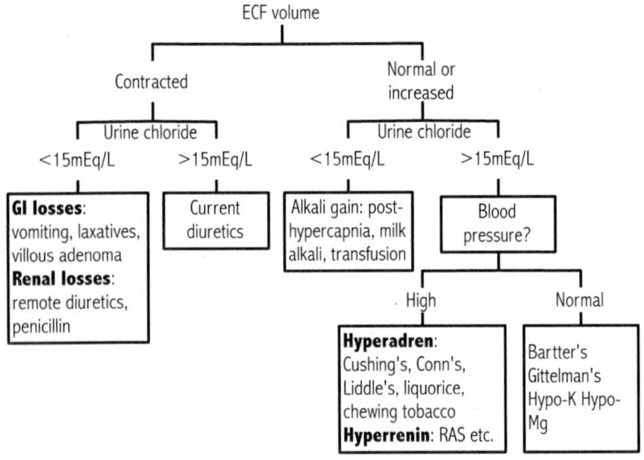

2. Treatment: Rx underlying causes

- If hypovolemic: give normal saline
- If hypervolemic: consider spironolactone or acetazolamide or KCl
- Correct hypokalemia and hypomagnesemia (if not corrected, metabolic alkalosis will persist)
- If pH > 7.7 causing seizure or ventricular arrhythmias, Call SMR/Staff. Dialysis or 0.1 N HCl @ 0.2mEq/kg/hr via central line.

RESPIRATORY ACIDOSIS

1. Elevated pCO2 (>45 mmHg) due to hypoventilation
2. Causes: "won't breathe"/ "can't breath" – See section on Hypercapnia for DDx
3. Renal response with increased serum HCO3 – acute vs. chronic (see above)
4. Treat underlying cause

RESPIRATORY ALKALOSIS

1. Low pCO2 (<35 mmHg) due to hyperventilation
2. Causes: PE, pneumonia, ILD, pain, anxiety, fever, sepsis, pregnancy
3. Renal response with decreased serum HCO3 – acute vs. chronic (see above)
4. Treat underlying cause

MIXED ACID-BASE PROBLEMS

1. Suspect mixed problems with extreme pH's (combination of two acidotic processes or two alkalotic processes) or normal pH (mixed acidosis and alkalosis; check HCO3, pCO2, pH and AG). Remember the pH does not usually normalise in sever primary processes; if pH in the normal range, think of two primary processes (e.g. primary metabolic acidosis and primary respiratory alkalosis).
2. Metabolic acidosis and respiratory acidosis:
   - Low HCO3 without appropriate hyperventilation, very low pH
   - Example: diarrhea and severe hypokalemia (latter causing resp muscle weakness)
3. Metabolic alkalosis and respiratory alkalosis:
   - High HCO3 without appropriate hypoventilation, very high pH
   - Example: hyperventilation and NG suction (ICU patients with increased ICP)
4. AG and non-AG metabolic acidosis
   - HCO3 fall is more than AG increase, pH low
   - Example: uremia with over-aggressive normal saline treatment; partially treated DKA
5. Metabolic acidosis and metabolic alkalosis:
   - HCO3 fall is less than AG increase, pH normal or near normal
   - Example: DKA patients with vomiting
6. Metabolic acidosis and respiratory alkalosis:
   - HCO3 fall with over compensation of low pCO2, pH normal or near normal Example: ASA intoxication
7. Metabolic alkalosis and respiratory acidosis:
   - HCO3 increase with over compensation of high pCO2, pH normal or near normal
   - Example: vomiting and decrease LOC (e.g. head trauma)
8. Metabolic acidosis, metabolic alkalosis, respiratory acidosis:
   - HCO3 fall is less than AG increase, normal or high pCO2 (i.e. no resp compensation)
   - Example: DKA patients with vomiting AND decreased LOC
9. Metabolic acidosis, metabolic alkalosis, respiratory alkalosis:
   - HCO3 fall is less than AG increase, very low pCO2 (i.e. resp over-compensation) Example: above DKA patients now on ventilator and being hyperventilated.

ACUTE MONOARTHRITIS

General Considerations:

- Acutely painful, swollen, erythematous, hot joint
- Most important emergent consideration is to rule out septic arthritis as this can lead to destruction of a joint in hours to days
- Can broadly subclassify into infectious, crystalline, inflammatory, degenerative, and traumatic

Differential Diagnosis:

- Infectious
  - Gonococcal; most common in young adults
  - Non-gonococcal (bacterial); *S. aureus* (60%), GAS (15%), *S. pneumonia* (3%), GNB (18%), anaerobes less common
  - Mycobacterial or fungal infection can occur in immunocompromised patients
- Crystalline arthropathy
  - Gout; severe pain and erythema, involves mostly distal lower extremity joints
  - CPPD; knee most common affected joint; wrist, shoulder, ankle also commonly affected
- Inflammatory
  - Seronegative arthropathy; reactive arthritis, psoriatic arthritis, IBD-associated arthritis
  - Seropositive disease; RA can rarely present as monoarthritis in early stages
- Trauma
  - Fracture, hemarthrosis, foreign body
- Non-joint
  - Osteomyelitis, AVN, tendinitis, bursitis

Approach:

- Rule out infectious etiology ie. septic arthritis
  - Fever, malaise, systemic signs of infectious including nausea, vomiting, chills
  - Risk factors: age > 80 (LR+3.5), diabetes (LR+2.7), recent joint surgery (LR+6.9) or prosthetic joint (LR+3.1)
  - All patients will need joint aspiration to rule out infection
- History should include precipitating events, including trauma, recent infectious history, travel history, risk factors for immunosuppression, family history, and sexual history if appropriate
- Physical exam should include examination of affected joint and surrounding joints and soft tissue structures
- Investigate for causes of arthritis as well as for coinciding systemic disease

Investigations:

- Joint aspiration
  - Highest yield investigation; reveals infection, level of inflammation
  - Send for: cell count and differential, culture and gram stain, crystal analysis

- Imaging
  - X-ray
    - Gout; in early disease only soft tissue swelling; late disease will reveal tophi and bony erosions
    - CPPD; chondrocalcinosis may be seen
    - Assess for fractures, signs of OA
  - Blood work
    - CBC, blood cultures
    - ESR/CRP; nonspecific but help distinguish inflammatory from non-inflammatory in setting of equivocal joint aspiration
    - RF +/- CCP if suspicion high for RA; however will not return overnight
    - Uric acid level; not as useful in acute setting as in chronic management of hyperuricemia
    - Other tests like HLA-B27, ANA, Lyme serology, etc not useful overnight

	Normal	Non-inflammatory	Inflammatory	Septic
Appearance	Clear, colourless	Fatty droplets – # Bloody – think hemarthrosis, tumour, or #		Cloudy
WBC/mm³	<200	<2,000	2,000-100,000	Usually >50,000
Neutrophils	<25%	<25%	>50%	>75%
Crystals			**Gout** – needle shaped, negatively birefringent **CPPD** – rhomboid shaped, positively birefringent	****presence does not rule out infection**
Gram Stain/ Culture	Negative	Negative	Negative	Positive (culture positive in 90% of cases)
Examples of Conditions		OA, AVN	CPPD, Gout, RA, seronegative	Gonococcal, non-gonococcal

Management:

1. **Septic Arthritis:** definitive management is surgical
   - Sepsis management if indicated; early antibiotics based on results of gram stain and culture, IV fluids, monitored setting
   - Contact surgical service for drainage/debridement
   - Rule out systemic illness including bacteremia (BCx)

2. **Gout/CPPD**
   - Primarily consists of NSAIDs OR colchicine OR oral/intraarticular steroid
     - Initiate as soon as possible after symptom onset; ideally within hours of onset

- NSAIDs; naproxen 500mg PO BID (ensure no peptic ulcer disease, normal renal function, caution with cardiovascular disease and heart failure) until acute episode resolves then discontinue
- Colchicine 1.2mg PO once, then 0.6mg PO BID, discontinue 2-3 days after episode resolves
- Corticosteroids
  - If NSAIDs/colchicine contraindicated
- Intraarticular; diagnosis must be certain to avoid injecting steroid into infected joint
- Oral; prednisone 30-50mg daily then taper over 7-10 days after episode resolves
- Urate-lowering therapy (allopurinol)
  - No role for initiating urate-lowering during a flare; however if on urate-lowering agent can continue throughout the episode.
  - Indications for urate lowering therapies: recurrent attacks, tophi, bony erosions, urate kidney stones.
- Lifestyle: avoid foods with high purine content (jred meats, sardines, shellfish), avoid drugs with hyperuremic effects (thiazides, alcohol)

3. Osteoarthritis

- NSAIDs; if comorbidities prevent oral NSAID use, can try topical agents (voltaren gel, capsaicin)
- Corticosteroids; intraarticular steroids generally reserved for longstanding osteoarthritis with an inflammatory component; little role for joint injection in the acute setting
- Nonpharmacologic; weight loss, physiotherapy, occupational therapy (walker, cane, bracing etc)
- Surgical management if refractory to above: joint debridement, osteotomy, total and or partial joint replacements)

POLYARTHRITIS/OLIGOARTHRITIS

General considerations:

- Common disorder with vast differential diagnosis
- Helpful to distinguish between arthralgia and arthritis (ie. presence or absence of synovitis)
- Otherwise the diagnosis primarily relies on thorough history taking
  - Morning stiffness (>1hr), pain at rest, improvement with movement, pain worse in morning point to inflammatory arthritis
  - Morning stiffness (<1/2 hr), pain with motion, relieved by rest, joint instability, bony enlargement/deformaties point to degenerative arthritis
  - Fever and other infectious symptoms point to infectious cause
  - Severe arthritis w/ significant swelling may point to crystalline or septic arthritis
  - Other careful history taking to elicit systemic signs of rheumatologic disease is important

Differential Diagnosis:

1. **Systemic Rheumatologic Disease**
   a. Rheumatoid arthritis (RA)- destructive autoimmune arthritis that often affects wrists, MCPs and PIPs symmetrically
   b. Lupus (SLE) – chronic inflammatory multi-system disease, consists of fever, joint pain, myalgias, malar/discoid rash, photosensitivity, serositis, neurological symptoms, oral ulcers, and multiorgan involvement (renal, hematologic disorders)
   c. Scleroderma –disorder with small-vessel vasculopathy and fibrosis; can be limited to the skin, or a diffuse involvement including kidneys, lungs, esophagus and skin. Often associated with CREST (Calcinosis, Raynaud's, Esophageal dysfunction, Sclerodactyly and Telangiectasia)
   d. Polymyositis/Dermatomyositis – inflammatory myositis leading to proximal muscle weakness
   e. Sjögren's – destruction of exocrine glands that produce tears and saliva leading to Sicca symptoms
   f. Adult Onset Still's Disease – a strongly inflammatory disorder associated with high fevers, salmon-coloured rash and leukocytosis
2. **Seronegative spondyloarthropathies**
   a. Reactive arthritis – generally one to three weeks following GI or GU infection (i.e. from chlamydia, shigella, salmonella, yersinia etc)
   b. Psoriatic arthritis – associated with skin manifestations, dactylitis
   c. Ankylosing Spondylitis – affects axial spine, sacroiliac joint and pelvis
   d. Inflammatory Bowel Disease – can have extra-intestinal manifestation of migratory arthritis
3. **Vasculitis**
   a. Granulomatosis with Polyangiitis (GPA/formerly Wegner's), Polyarteritis Nodosa (PAN), Giant Cell Arteritis (GCA), Henoch-Schonlein Purpura (HSP)
4. **Infectious**
   a. Bacterial – less commonly polyarthritis rather than monoarthritis. More likely polyarthritis if gonococcal. Consider group A streptococci (rheumatic fever)
   b. TB
   c. Viral – parvovirus, enterovirus, adenovirus, EBV, CMV, rubella, mumps, hepatitis B/C, varicella, HIV

d. Spirochetes - Lyme – can also present as monoarthritis
e. Fungal

5. **Crystal Joint Arthropathies**
   a. Gout
   b. Calcium Pyrophosphate Dihydrate Crystals (CPPD)
   c. Other - Calcium Oxalate crystals or apatite crystals
6. **Osteoarthritis** – tends to affect DIP, PIP and 1st CMC. Morning stiffness, if present, lasts < 1 hour.
7. **Other**
   d. Serum sickness – after transfusion, associated with lymphadenopathy, fever and rash
   e. Malignancy/Multiple myeloma
   f. Amyloidosis

Physical Examination:

- Physical exam guided by differential diagnosis
- Infectious
  - **Always rule out septic arthritis first**; exquisitely tender to any ROM, significant erythema, pain, effusion
  - Assess for lymphadenopathy, splenomegaly, rash, bites, consider pelvic exam if suspicious
- Assess for systemic findings of rheumatologic disease
  - *Rheumatoid arthritis* – nodules, interstitial lung disease, secondary sjögrens, scleritis and rheumatoid vasculitis, mononeuritis multiplex
  - *SLE* – malar rash, discoid rash, photosensitivity, serositis (pleuritis/pericarditis), oral ulcers, alopecia
  - *Seronegative spondyloarthropathies* – iritis, conjunctivitis, back pain, sacroilitis, pyoderma gangrenosum
  - *Dermatomyositis* – Heliotropic rash, Gottren's papules
  - *Scleroderma* – skin thickening
  - *Rheumatic fever* – chorea, erythema marginatum rheumaticum
  - *Vasculitis* – saddle nose deformity, palpable purpura, livedo reticularis, mononeuritis multiplex, ischemia

Investigations:

1. **Autoimmune**
   a. RA – RF +/- anti-CCP, X-Rays reveal joint space narrowing, erosions, periarticular osteopenia
   b. SLE – ANA/ENA, anti-dSDNA, C3/C4
   c. Sjögrens – Anti-SS-A (anti-Ro), anti-SS-B (anti-La)
   d. Scleroderma – ANA/ENA
   e. Seronegative arthritis – HLA-B27
2. **Vasculitis** – ESR, biopsy of affected area, c-ANCA (Wegner's)
3. **Infectious** – for bacterial see section on monoarthritis.
   a. Viral – consider EBV/CMV/Parvovirus serology; rule out HIV/Hepatitis if suspicious
4. **Crystal Arthropathy** – joint aspiration if affecting large joint
5. **Osteoarthritis** – X-Ray findings include joint space narrowing and osteophytes

Autoantibody and Rheumatic Diseases:

Rheumatic Disease	Autoantibody
Seropositive Rheumatic Diseases	
Rheumatoid Arthritis (RA)	RF, Anti-CCP
Systemic Lupus Erythematosus (SLE)	ANA, Anti-dsDNA, anti-Sm, APLA
Scleroderma	Anti-topoisomerase I (anti-Scl-70) (diffuse systemic sclerosis), Anti-centromere (CREST syndrome)
Sjogren's Syndrome	Anti-Ro (Anti-SS-A), Anti-La (Anti-SS-B)
Myositis/anti-synthetase syndrome	Jo-1
Seroneative Rheumatic Diseases	
Seronegative arthritis	HLA-B27
Vasculitis	
Granulomatosis with polyangitis (previously Wegener's)	c-anca
Eosinophilic granulomatosis with polyangitis (Churg-Strauss)	p-anca
Non-anca vasculitis	ESR

Management:

1. **Systemic Rheumatologic Disease**
   a. Rheumatoid arthritis (RA) – first line management involves DMARD (methotrexate) +/- steroids.
   b. Lupus (SLE) – NSAIDs, hydroxychloroquine, steroids, methotrexate
2. **Seronegative Arthritis** – Consider NSAIDS, DMARDS (methotrexate, leflunomide), biologics
3. **Vasculitis** – induction therapy with pulse-dose steroids & cyclophosphamide
4. **Infection**
   a. For bacterial and gonococcal see *Infectious Disease* section
   b. For arthritis related to self-limiting viral infections, usually requires symptomatic treatment only with analgesics or NSAIDs
5. **Crystal Joint Arthropathies** – See section on monoarthritis
6. **Osteoarthritis** – see section on monoarthritis

Definition:

- Macule – flat lesion <1cm
- Plaque – flat lesion >1cm
- Papule – raised lesion <1cm
- Nodule – raised lesion >1cm
- Pustule – lesions containing pus
- Vesicle – fluid filled raised lesion <1cm
- Bullae – fluid filled raised lesion >1cm
- Petechiae – non blanchable red/brown <1cm lesion, associated with thrombocytopenia

Differential Diagnosis:

1. **Infection** – viral exanthem, scalded skin syndrome, necrotizing fasciitis, endocarditis, scabies
2. **Allergy** – important to rule out anaphylaxis; new medications, contact dermatitis
3. **Bullous Rash** – Bullous pemphigoid, pemphigus vulgaris, burns, contact dermatitis, necrotizing fasciitis, varicella, MRSA
4. **Cutaneous vasculitis** – leukocytoclastic vascultitis (associated with drug exposure, serum sickness, viral infection, bacterial endocarditis, hepatitis/cryoglobulins, IBD, collagen vascular diseases, systemic vasculitis or malignancy)
5. **Mucosal involvement** – think erythema multiforme (EM) major, toxic epidermal necrosis (TEN), Stevens-Johnson Syndrome (SJS), pemphigus vulgaris. Toxic shock syndrome. In children, consider Kawasaki disease and Scarlet fever.
6. **Petechiae**
   - **Non-palpable-** Thrombotic thrombocytopenic purpura (TTP), DIC, ITP
   - **Palpable**- HSP, endocarditis, autoimmune vasculitis, meningococcemia, disseminated gonococcal infection, RMSF
7. **Plaque lesion**- Psoriasis
8. **Eczema**
9. **Other-** dermatitis herpetiformis (associated with Celiac disease), pityriasis rosea

Approach:

- History and physical exam guided by level of acuity and by differential diagnosis
- For the acutely unwell patient
  - Severe, blistering and sloughing rash should prompt consideration of **TEN** or **SJS**
  - Fever, hypotension, and appropriate history and physical findings can point to **endocarditis** or **meningitis/meningiococcemia**
  - Any combination of hypotension, tongue swelling, wheezing and respiratory distress with a drug exposure should be managed as **anaphylaxis** until proven otherwise
- For less acute presentations
  - Consider coexisting conditions including infection, inflammatory or autoimmune disease, drug or allergen exposures, travel and occupational history, sexual history if appropriate
  - Investigations guided by likely causative conditions or exposures
- Diagnosis often hinges on history and physical exam findings, which guide appropriate investigations for likely causes
- Management of the underlying cause, administer treatment concurrently along with diagnostic tests if severe illness

SMOKING CESSATION

Assessing Use/Exposure:

- Smoking is the leading preventable cause of mortality
- Many want to quit, many have multiple attempts (avg. 6-7 attempts), very few remain quit after 12 mos
- Always ask about use (past, present, 2nd hand exposure) and product (cigarette, pipe, chewing tobacco…)
- Obtain a quit history- what has worked, what hasn't, how many times, why did they restart?

5 A's Algorithm:

- **Ask**- always ask (as described above)
- **Advise**- encourage to quit, inform about risks of smoking/tobacco exposure; personalize your advice—tie tobacco use with current health/illness, cost, impact on children/others in household
  - Stopping smoking can be difficult, effective treatment options for tobacco dependence are available, we are here to help you quit
- **Assess**- assess willingness to quit in the next 30 days- if so, move on to assistance; if not, provide motivational intervention (5R's: relevance to pt, risks of smoking, rewards for not smoking, roadblocks to quitting, repetition of motivational intervention) and reassess reasons for smoking
- **Assist**- provide aid for the patient to quit and set a quit date
- **Arrange**- arrange for several follow-up opportunities (ex. With Family Physician); should occur within 1 week of quit attempt
  - Reward success, review any tobacco use and understand why it happened (learn from lapses)
  - Identify problems encountered and anticipate challenges

Cessation Aids:

- Non-pharmacotherapy
  - Inform about what to expect when quitting, CBT, trigger identification/avoidance (ex. EtOH, locations)
  - Behavioural/cognitive distraction (doodling, thinking about other things- to-do lists), chew gum, drink water, visualization
  - Stress management/relaxation strategies
  - Counselling- one-on-one, group settings (Family Doctor may have access to this)
  - Canadian Cancer Society's Smoker's Helpline: 1877-513-5333, or Texting: iQUIT to 123456, Hamilton Smokers Helpline
  - Self-help resources, phone apps
- Pharmacotherapy
  - First Line: NRT (transdermal patch, gum, lozenge, inhaler, nasal spray), varenicline, bupropion
    - NRT goal: reduce nicotine withdrawal symptoms during quit attempt (often used in inpatients)
    - Patch can give baseline nicotine, gum/lozenge can control cravings PRN

        - NRT increases quit rates twofold
        - HHS order sets for nicotine replacement- initial nicotine replacement amount depends on the amount of nicotine used daily
        - NRT used in combination is safe, and roughly as effective as using varenicline
    - Varenicline: partial agonist of nicotinic Ach receptor the reinforces effects of nicotine and leads to dependence
        - Reduces symptoms of nicotine withdrawal and reduces rewards to cigarette smoking
        - Increases quit rates by over twofold
        - Requires renal dosing; careful psych history prior to using and stop if any mood/behavioural changes, currently unstable psych status or history of suicidal ideation
        - Start at 0.5mg PO daily x3d, then 0.5mg PO BID x4d, then 1mg PO BID x12 weeks; start one week before the quit date
    - Bupropion: enhances CNS noradrenergic and dopaminergic release
        - Improves quit rates by approximately twofold
        - 150mg PO daily x3d, then 150mg PO BID x12 weeks; start one week before the quit date
        - Reduces seizure threshold, therefore contraindicated in seizure disorder
    - Other pharmacotherapies have been tried, but there is no good evidence to support them: nortriptyline, cytisine, SSRIs, E-cigarettes
- If a patient fails therapy, ask:
    - Were they taking the medication properly? (ex. Not chewing the gum too quickly/not parking)
    - Were there too many side effects?
    - Did it fail to reduce withdrawal symptoms?

Resources in Hamilton
- Hamilton Smokers Helpline (905)540-5566

NUMBERS OF COMMUNITY AGENCIES

Shelters- Adult Male
- Good Shepard (135 Mary St) (905)528-9109
- Mission Services (325 Jams St N) (905)528-7635
- Salvation Army (94 York Blvd) (905)527-1444

Shelters- Adult Female
- Inasmuch House (Women + children leaving abusive situations) (905)529-8149
- Interval House of Hamilton (Women leaving abusive situations) (905)387-8881
- Martha House (Women + children leaving abusive situations) (905)523-8895
- Mary's Place (905)540-8000
- Mountain View Program (Aboriginal Women, East Mountain) (905)664-1114
- WomanKind (West Hamilton, near McMaster) (905)521-9591

Shelters- Families

- Families First (905)528-9442
- Good Shepherd Family Centre (905)528-9442

Shelters- Youth (21 and under)

- Notre Dame (14 Cannon St W) (905)308-8090

DETOX

- Discovery House and Berringer House (325 James St N) (905)528-7635
- Suntrac (905)528-0389
- Mens Addiction Services of Hamilton (905)527-9264
- ADGS (Hunter St) (905)546-3606
- Alternatives For Youth (905)527-4469
- ADAPT (Burlington/Oakville area residents) (905)639-6537
- Wayside House- Hamilton (Charlton St) (905)528-8969
- Wayside House- Niagara (St. Catharines) (905)684-9248

Managed Alcohol Facilities

- Wesley Urban Ministries – Special Care Unit (905)318-6903

Additional Support Services in Hamilton

- Barrett Centre (Mental Health Crisis Stabilization Service) (905)529-7878
- Wesley Urban Ministries (various services incl. harm reduction) (905)528-5640
- Hamilton Urban Core (various services incl. help with ID) (905)522-3233
- Housing Help Centre (905)526-8100
- YMCA (Accommodation- (905)317-4912) (905)529-7102

Crisis Lines

- AA Hotline (905)522-8392
- CA Hotline (905)544-7991
- NA Hotline (905)522-0332
- COAST (Community Outreach and Support Team) (905)972-8338
- Connex Ontario (Drugs and Alcohol Helpline) DART 1800-565-8603

Health Clinics/Health Outreach

- HamSMART (Hamilton Social Medicine Response Team) (905)521-2100 ext. 42471
  - For patients who have difficulty accessing care through traditional means, who are affected by poverty, homelessness, addiction, or are frail elderly and have difficulty with outpatient follow up or care
  - For patients who want to start methadone treatments as inpatients
  - Fill out online referral form and fax to (905)575-7320
  - www.hamsmart.com
- Shelter Health Network (905)667-0474

Methadone

- If a person expresses interest in starting methadone or suboxone, a referral for inpatient start of methadone or suboxone can be made to HamSMART
- Community Referrals - Hamilton Clinic (905)523-4567

PROCEDURES

GENERAL PREPARATION

1) Obtain informed consent
2) Ensure no contraindications present (check INR, aPTT, platelets, surface exam for cellulitis/tissue infection)
3) Obtain kits, local anesthetic, assistance, sterile materials, additional sample/drainage/culture bottles
4) Ensure patient is in proper position, with bed and materials in comfortable location for individual performing the procedure

PARACENTESIS

Indication:

1. Establish etiology of new onset ascites
2. Rule out SBP
3. Therapeutic-large volume paracentesis for tense ascites causing dyspnea or abdominal discomfort

Contraindications:

1. DIC
2. Coagulopathy and thrombocytopenia (relative contraindication)
3. Do not place needle through sites of: Infection, engorged subcutaneous vessels, surgical scars, or hematomas
4. Caution should be done if pregnant, SBO, organomegaly or adhesions

Equipment:

- Paracentesis/thoracentesis kits *will have the needle and catheter, sterile drapes, gauze
- Sterile gloves
- Face shield
- Chlorhexidine wipes x3
- 1 or 2% lido without epi
- 10cc syringe
- Blunt Red capped needle (16G)
- Black capped needle (22G)
- 60cc syringe
- Large evacuated bottles
- Sterile dressing
- Culture bottles
- U/S

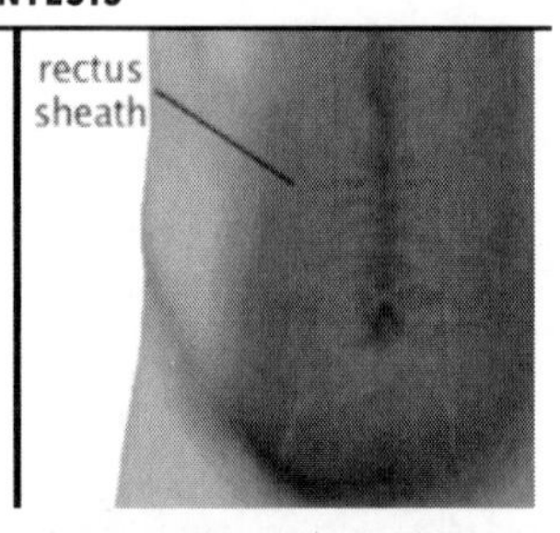

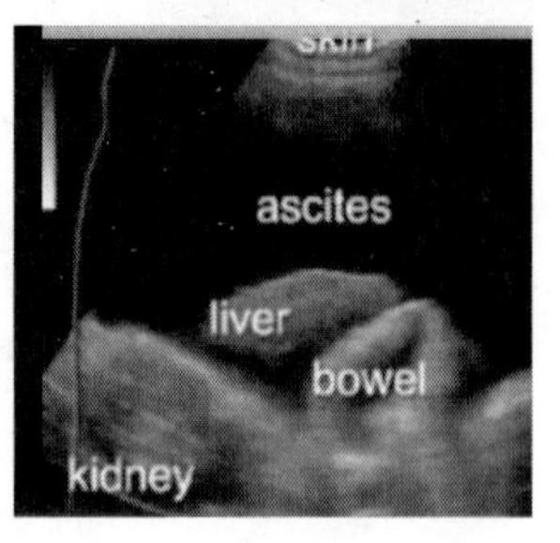

Complications:

1. 0.2% of serious complications (intraabdominal injury and puncture of inferior epigastric artery)
2. Infection, hematoma, continued leakage, hypotension

Position of patient/landmarking:

1. Patient should be supine, arms at side, 1 pillow under head for comfort is ok
2. Landmarking: Midline, 2 cm below umbilicus (advantage: devoid of blood vessels) or left/right lateral, 2-4 cm medial and cephalad to ASIS (advantage: more ascites, less sub cutaneous fat)
   - If using a lateral approach, make sure to be lateral of the rectus sheath to avoid hitting the inferior epigastric artery (picture right)
   - If using subumbilical, make sure to empty the bladder

Procedure:

1. Use clinical exam to determine presence of ascites and U/S exam if it is available to confirm your location
2. If using U/S, find an appropriate pocket of fluid (right)
3. Mark the site using a needle cap to dent the skin
4. Bring the height of the bed to a comfortable elevation, get all of your supplies ready and opened
5. Glove yourself
6. Clean the area that you marked x3, using chlorhexidine with progressively widening concentric circles
7. Place the sterile drapes on your patient to create a sterile field
8. Have someone else hold the lidocaine and draw up using the red capped needle into the 10cc syringe
9. Change to the black capped needle
10. Make a small wheal under the skin of lido at the mark that you created *recall always draw back before injecting lidocaine to make sure you aren't injecting into a blood vessel*
11. Then, going in at about 45 degrees, slowly inject lidocaine→advance the needle a further 2-3mm→pull back on the syringe→repeat until you see ascitic fluid being drawn up into your syringe. At that point in time inject the remaining lidocaine to anesthetize the parietal peritoneum.
12. **IT IS IMPORTANT TO NOTE THE DIRECTION OF THE NEEDLE AND THE DEPTH OF THE NEEDLE→use this needle track as a guide!**
13. Withdraw your lidocaine needle/syringe
14. Take your 60cc syringe and place it on the thoracentesis needle/catheter
15. Make sure that the catheter and needle glide easily past one another
16. Insert thoracentesis needle using angular (45 degrees to skin) or Z (perpendicular through epidermis then move skin and needle 2 cm caudal and insert through peritoneum) techniques to reduce chance of continued leak and infection post procedure
17. Proceed with dominant hand on the syringe and non-dominant hand on skin/supporting the needle, advance at 2-3 mm increments with continuous negative pressure on syringe. When ascitic fluid returned, advance the needle 1-2mm further, then stop advancing needle, advance catheter and remove needle.
18. Connect stopcock, fully collect fluid in the 60cc syringe, attach the evacuated containers to the stopcock and then remove the remaining fluid
19. If over 3 litres removed, consider replacement with albumin (6-8 gm/litre removed)
20. Place a sterile bandage on the site
21. Check the patient's blood pressure

LUMBAR PUNCTURE

Indications:
- Spinal/epidural anesthesia
- Diagnostic-meningitis, leukemia, metabolic, inflammatory (GBS)
- Therapeutic-antibiotics and chemo

Contraindications:
- Cardioresp compromise
- Avoid in patients with focal neuro signs, ?herniation, ?Increased ICP
  - o Do a CT if you have any suspicions
- Coagulopathy/on anticoagulant
- Suspected Epidural abscess
- If had a prev spinal surgery→ consult IR to do the tap

Equipment
- A lumbar puncture kit→ contains the spinal needle (in adults that is about 8.9cm in length)
- Sterile drapes
- Sterile gloves
- Sterile gown
- Face shield
- 1 or 2% lido without epi
- 10cc syringe
- Blunt Red capped needle (18G)
- Black capped needle (22G)

Complications:
- Herniation, cardiorespiratory compromise, headache, bleeding, infection, subarachnoid epidermal cyst, pain, CSF leakage

Positioning of patient/landmarking:
1. Patient should be either sitting up slumped over a table or in the lateral recumbent position with their back flexed to widen the space between the spinout processes (above right)
2. Landmarking: Palpate the superior aspect of the iliac spines (will intersect L4 spinous process) –plan to insert the needle between the interspace of L3-L4 or L4-L5 spaces as this is below the conus medularis (L1-L2)
3. Avoid overlying skin infections
4. Mark location prior to local aesthetic and cleaning (can obscure landmarks)

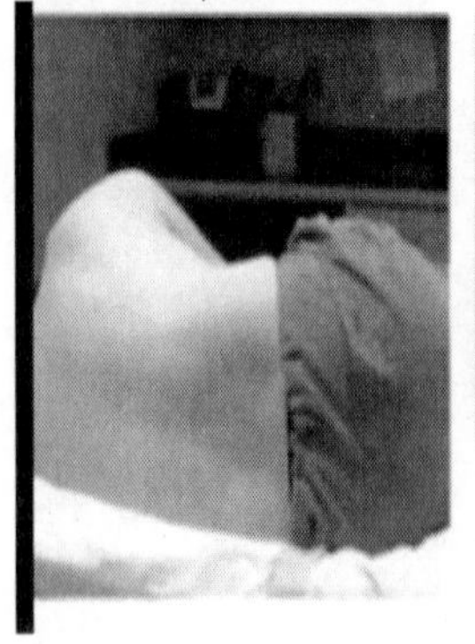

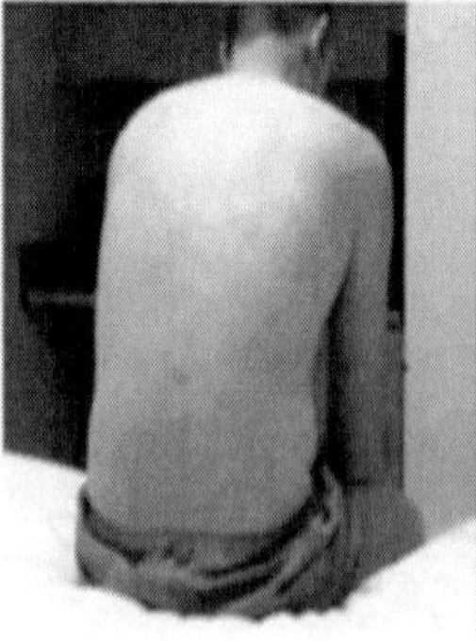

Procedure:

1. Bring bed height to comfortable elevation, get all the supplies ready and opened
2. Mark the site using a needle cap to dent the skin
3. Glove/gown yourself
4. Clean the area that you marked x3, using chlorhexidine
5. Place the sterile drapes on your patient to create a sterile field
6. Have someone else hold the lidocaine and you draw up your lido using the red capped needle into the 10cc syringe
7. Change to the black capped needle
8. Apply a small wheal under the skin at your landmark *recall always draw back before injecting lido to make sure you aren't injecting into a blood vessel
9. Then enter the skin at 2-3mm increments, aspirating and injecting lido in a fan-like distribution
10. Withdraw your lidocaine needle/syringe
11. Re-landmark, and hold your LP needle with the stylet in place at the superior aspect of inferior spinous process.
12. Holding the needle parallel to the bed, with the direction of the needle pointing towards the umbilicus, and bevel facing cephaled (to spread rather than tear dura thus minimize post LP headache), enter the skin
13. Feel for the tissues that are being passed through (skin, subcutaneous tissue, supraspinous ligament, interspinous ligament, ligamentum flavum, epidural space including the internal vertebral venous plexus, dura, arachnoid, and then between the nerve roots of the cauda equina → will feel a "pop" when passing through the ligamentum flavum)
14. Once feel the pop, advance needle at 2mm increments, followed by removing the stylet to look for CSF flow
15. If no flow try to rotate needle 90 degrees and recheck for flow
16. If still no flow, withdraw needle to sub cut tissue and redirect needle
17. Measure opening pressure if in lateral decubitus position-attach a stopcock to the needle and then attach the manometer to that-close the stopcock to the outside and watch the csf rise
    - If >25→ increased ICP
18. Once that is done, turn the stop cock to the patient and using the fluid in the manometer, fill each of the 4 tubes provided in the kit with 3-4 ml/tube *important to fill them in order tube 1=1st, tube 2=2nd, etc*
19. If you need more fluid, turn the stopcock to the manometer and fill the remaining tubes
20. Replace stylet once you have filled all your tubes
21. Remove needle
22. Place a sterile bandage on the patient

What to send the tubes for:
Tube 1: Cell count and differential
Tube 2: Gram stain, bacterial and viral cultures
Tube 3: Glucose, protein, protein electrophoresis
Tube 4: Cell count and differential, special tests including cytology, AFB cultures, etc

THORACENTESIS

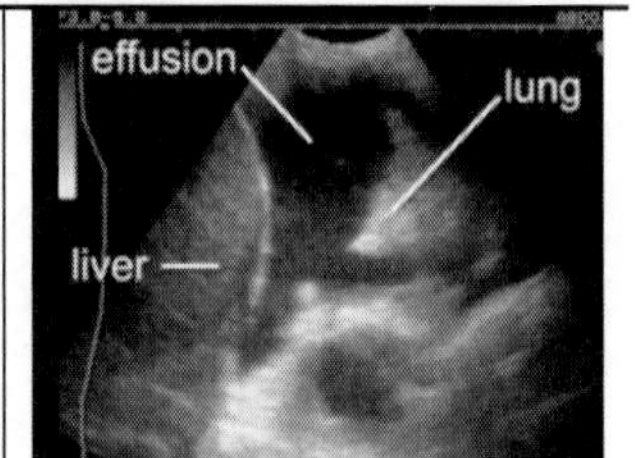

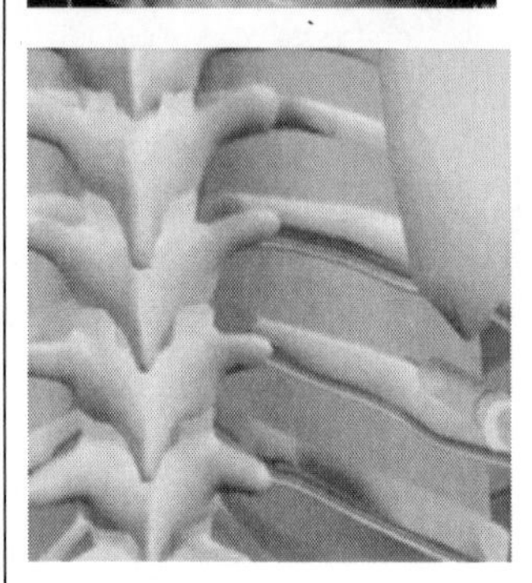

Indications:
- Diagnostic: find the cause of pleural effusion
- Therapeutic for respiratory compromise

Contraindications:
- Relative: coagulopathy (ok if INR<1.5), thrombocytopenia
- Avoid if hemodynamic or respiratory compromise or mechanically ventilated unless effusion is the cause of the respiratory insufficiency
- Uncooperative patient
- Do not put needle through sites of cutaneous infections (i.e. herpes zoster or cellulitis)

Equipment:
- Thoracentesis kits *will have the needle and catheter, sterile drapes, gauze, 3 way stopcock*
- Sterile glove
- Face shield
- Chlorhexidine wipes x3
- 1 or 2% lido without epi
- 10cc syringe
- Blunt Red capped needle (16G)
- Black capped needle (22G)
- 60cc syringe
- Large evacuated bottles
- Sterile dressing
- Culture bottles
- U/S

Complications:

- Pneumothorax, pain, coughing, bleeding, infection, post expansion pulmonary edema, air embolism and intra-abdominal injury

Position/landmark

1. Patient should be sitting up, leaning over bedside table (decubitus can also be used)
2. Landmark effusion (dullness to percussion, decreased breath sounds and tactile fremitus) then mark 2-3 intercostal spaces below the upper border of the fluid and then 5-10 cm lateral to the spine, but not below the ninth rib (minimizes intra-abdominal injury)
3. If small effusion, use ultrasound guidance/marking (see above R picture)

Procedure:

1. Bring the height of the bed to a comfortable elevation with your patient positioned appropriately
2. Mark the site using a needle cap to dent the skin
3. Prepare all the materials what you will need and open them in a sterile manner
4. Glove/gown yourself
5. Clean the area that you marked x3, using chlorhexidine
6. Place the sterile drapes on your patient to create a sterile field
7. Have someone else hold the lidocaine and you draw up your lido using the red capped needle into the 10cc syringe
8. Change to the black capped needle
9. Apply a small wheal under the skin at your landmark *recall always draw back before injecting lidocaine to make sure you aren't injecting into a blood vessel
10. Then, continue into the subcutaneous tissue along the superior edge of inferior rib to avoid the neurovascular bundle that sits on the inferior aspect of each rib. Advance the needle 2-3mm at a time while aspirating, then inject the freezing. Continue inserting the needle until you aspirate the pleural effusion, then inject lidocaine to freeze the parietal pleura
11. **NOTE THE TRAJECTORY OF THE NEEDLE!**
12. Repeat above with thoracentesis needle, advancing slowly with negative pressure on syringe.
13. Once fluid returns, advance catheter and remove the needle while having the patient hum and then quickly cover the catheter with your finger once the needle is completely withdrawn.
14. Attach to three-way stopcock
15. Close the stopcock to the atmosphere and draw up fluid in your 60cc syringe
16. Then, if more fluid is to be removed, attach the evacuation bottles and close the stopcock to the syringe.
17. Remove catheter during the patient's end-expiration to prevent pneumothorax
18. Drain up to maximum of 1.5L of fluid to prevent post-expansion pulmonary edema

Central Venous Catheter (see "SHOCK")

ICU DATA CARD

	CALCULATION	NORMAL RANGE
Cardiac Index (CI)	$\frac{SV \times HR}{BSA}$	2.5-4.2 $L/min/m^2$
Mean Arterial Pressure	$DBP + \frac{SBP - DBP}{3}$	80-100 mmHg
Mean Pulmonary Artery Pressure (PAP)	$PAD + \frac{PAS - PAD}{3}$	11-18 mmHg
Systematic Vascular Resistance Index (SVRI)	$\frac{(MAP - CVP) \times 79.9}{CO\text{-}m^2}$	1970-2390 $dynes.sec.cm^3/m^2$
Pulmonary Vascular Resistance Index (PVRI)	$\frac{MPAP - PCWP \times 79.9}{CO\text{-}m^2}$	225-315 dynes.sec. cm^3/m^2
Arterial O_2 Content = (CaO_2)*	$(Hgb \times 1.34)\ SaO_2 + (PaO_2 \times 0.0031)$	16-22mls O_2/dl blood (vol %)
Mixed Venous O_2 content (CVO_2)*	$(Hgb \times 1.34)\ SvO_2 + (PvO_2 \times 0.0031)$	12-17 mls O_2/dl blood (vol %)
AVO_2 difference ($C(a\text{-}v)O_2$)*	$CaO_2 - CvO_2 = (Hgb \times 1.34) \times (SaO_2 - SvO_2)$	3.5-5.5 ml O_2/dl blood (vol %)
Oxygen Delivery (DO_2)	$CaO_2 \times CO \times 10$	700-1400 ml/min
Oxygen Consumption (VO_2)	$(CaO_2 - CvO_2) \times CO \times 10$	250 ml/min
Left Ventricular Stroke Work Index (LVSWI)	$SV(MAP - PCWP) \times 0.136/m^2$	35-60g/beat/m^2
Alveolar O_2 Pressure (PAO_2)	$(PB - PH_2O) \times FIO_2 - \frac{PaCO_2}{0.8}$	Variable
Alveolar-Arterial O_2 Gradient (A-a gradient)	$PAO_2 - PaO_2$	Room air: 2-25 mmHg
Shunt Fraction (QS/QT)	$\frac{CcO_2 - CaO_2}{CcO_2 - CvO_2}$	≤ 0.05
Proportion Dead Space (VD/VT)	$\frac{PaCO_2 - PECO_2}{PaCO_2}$	0.20
Cerebral Perfusion Pressure (CPP)	MAP-ICP	70-90 mmHG

* hemoglobin in g/dl units

	NORMAL RANGE
RA, CVP	2-10mmHg
RV	15-30/0-5mmHg
PA	15-30/8-15mmHg
PCWP	5-16mmHg

INTUBATION AND EXTUBATION GUIDELINES

Intubation Guidelines

(individual assessment required)

RR>40/min

$PaCO_2$ >60mmHg with respiratory acidosis

PaO_2 <60mmHg on 100% O_2

Airway protection

Bronchopulmonary toilet

High spinal cord injury

Extubation Guidelines

(individual assessment required)

MIP <-20cmH20

RR<30

TV>5cc/kg

VC>10cc/kg

PaO_2>65mmHg on FIO_2 40%

VE<10L/min

Able to protect airway

QS/QT<20%

VD/VT<60%

Glomerular Filtration Rate (GFR)

$$= \frac{\text{weight (kg) (140-age)}}{\text{Plasma Creatinine (umol/L)} \times 0.8}$$

DETERMINING PRERENAL VS. ATN IN AKI

	Prerenal	**Renal (ATN)**
Uosm (mosm/L)	>500	~300 (serum Osm)
UNa (mEq/L)	<20	>40
Fractional Excretion of Na	<1%	>2%
Fractional Excretion of Urea	<35%	>35%
U specific gravity	>1.015	<1.015
Urea/Cr relationship	If urea × 10 >Cr, then prerenal	

STANDARDIZED MINI-MENTAL STATUS EXAM (SMMSE)

Section 1

1.	Year	☐	1	6. Country	☐	1
2.	Season	☐	1	7. Province/State/ County	☐	1
3.	Month	☐	1	8. City/Town	☐	1
4.	Today's Date	☐	1	9. Place	☐	1
5.	Day of the Week	☐	1	10. Floor of Building	☐	1

Section 2

11.	Word 1	☐	1	16. No if's, and's or but's	☐	1
	Word 2	☐	1	17. Subject closes eyes	☐	1
	Word 3	☐	1	18.Takes paper in correct hand	☐	1
12.	DLROW or Serial Sevens	☐	5	Folds it in half	☐	1
13.	Word 1	☐	1	Puts it on the floor	☐	1
	Word 2	☐	1	19. Sentence	☐	1
	Word 3	☐	1	20. 4-sided figure in 2 five-sided figures	☐	1
14.	Wristwatch	☐	1			
15.	Pencil	☐	1	Total Score	☐	30

Total SMMSE Scores:

24 – 30 Normal Range
20 – 24 Mild Cognitive Impairment
10 – 20 Moderate Cognitive Impairment
0 – 10 Severe Cognitive Impairment

Diagnosis:

		Reference Standard	
		Positive	Negative
Test Result	Positive	a	b
	Negative	c	d

Sensitivity: proportion of people with the target disorder in whom a test result is positive
Sensitivity = a / (a + c)
SNOUT: 'Rule OUT if SENSITIVE test'

Specificity: proportion of people without the target disorder in whom a rest result is negative
Specificity = d / (b + d)
SPIN: 'Rule IN if SPECIFIC test'

Likelihood Ratio (LR): indicates by how much a given diagnostic test result will raise or lower the pretest probability of the target disorder.
LR for Positive Test (LR+) = [a / (a + c)] / [b / (b + d)]
LR for Negative Test (LR-) =[c / (a + c)] / [d / (b + d)]

LR = 1.0, pre-test probability = post-test
LR > 1.0, post-test probability > pre-test probability
LR < 1.0, post-test probability < pre-test probability

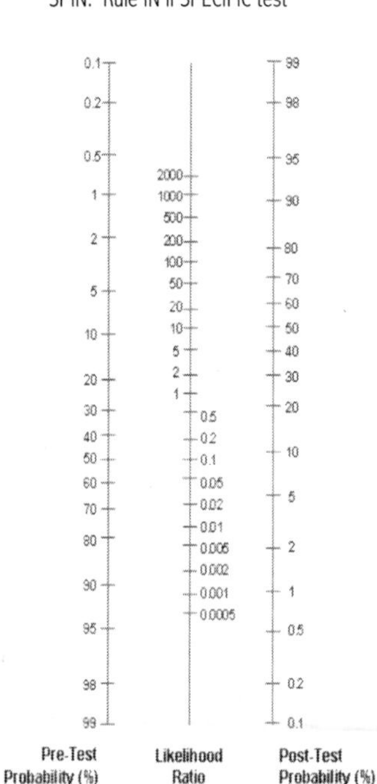

Pre-test probability: estimate based on clinical assessment
Post-test probability: based on pre-test probability and likelihood ratio nomogram

Positive Predictive Value (PPV): proportion of people with a positive test result with the target disorder
PPV = a / (a + b)

Negative Predictive Value (NPV): proportion of people with a negative test result without the target disorder

Therapy:

		Outcome	
		Present	Absent
Exposure/ Treatment	Present	a	b
	Absent	c	d

Controlled Event Rate (CER) = c / (c + d)

Experimental Event Rate (EER) = a / (a + b)

Relative Risk (RR): proportion of the original risk that is still present when patients receive the exposure.
RR = EER/CER

Relative Risk Reduction (RRR): an estimate of the proportion of baseline risk that is removed by the exposure.
RRR = 1-RR

Absolute Risk Reduction (ARR): proportion of patients spared the outcome in the exposed versus unexposed.
ARR = CER – EER

Number Needed to Treat (NNT): the number of patients who need to be treated over a specific period of time to prevent one target outcome.
NNT = 1/ARR

Odds Ratio (OR): proportion of patients with the target outcome divided by the proportion without the target outcome.
OR = (a/b)/(c/d) = ad/cb

USEFUL LIKELIHOOD RATIOS: HX, PE, AND IX

MEDICAL CONDITION	DIAGNOSTIC TEST	Sn (%)	Sp (%)	LR (+)	LR (−)
Acute Pancreatitis ^	- Serum amylase	82	91	9.1	0.2
	- Serum lipase	94	96	24	0.06
Alcohol Problem*	- CAGE questionnaire				
	Score 0			0.14	
	Score 1			1.5	
	Score 2			4.5	
	Score 3			13.3	
	Score 4			100	
Anemia - Iron Deficiency ^	-Ferritin				
	<15 ug/L	59	99	59	0.41
	<25 ug/L	73	98	37	0.28
	<100 ug/L	94	71	3.2	0.08
	- Transferrin saturation <10%	49	88	4.1	0.58
Ascites*	- Hx of increased girth	87	77	4.16	0.17
	- Bulging flanks	72 - 93	44-70	1.4-2.4	0.1-0.5
	- Fluid wave	50 - 80	82-92	2.8-9.6'	0.2-0.6
	- Shifting dullness	60 - 88	56-90	1.9-5.8	0.2-0.5
Aortic Stenosis*	- Slow rising carotid pulse			2.8-130	0.18-0.73
	- Any murmur present				
	- Systolic murmur			2.4	0
	Mid-peaking			8.0	0.13
	Late-peaking			101	0.31
	Rad. to right carotid			1.4 - 1.5	0.05 - 0.1
	- Quiet or absent S2			3.1 - 50	0.36 -
	- Apical carotid delay			¥	0.45
					0.05
Coronary Artery Disease ^	- Exercise ECG				
	ST depressions				
	>= 0.5mm	86	77	3.7	0.18
	>= 1.0mm	65	89	5.9	0.39
	- Dobutamine myocardial perfusion imaging	91	86	6.5	0.10
Deep Vein Thrombosis ^	- Venography	100	100	¥	0
	- Doppler ultrasound	95	95	19	0.05
Dementia ^	- MMSE <24	87	82	4.8	0.16
Meningitis* Bacterial or Viral	- Headache	50	15 -		
	- Fever	85	50		
	- Neck stiffness	70	45		
	- Altered mental status	67			
	- Kernig sign	9			
	- Brudzinski	15	100		
	- Jolt accentuation	97	100		

^ Adapted from Diagnostic Strategies for Common Medical Problems, 2E ACP 1999.
* Adapted from JAMA "Rational Clinical Exam" series.

CONSULT NOTE TEMPLATE

1. Date/Time:
2. Attending Name:
3. ID (Identification age, where from):
4. RFR (Reason for Referral):
5. PMHx (Past Medical History):
6. Meds (Medications):
7. Allergies: drugs (reaction)
8. HPI (History of Presenting Illness): including prior to hospital and in hospital
9. SHx (Social History):
10. FHx (Family History):
11. ROS (Review of Systems):
12. O/E (On Examination):
13. Investigations: including bloodwork, x-rays, EKG's
14. Impression: (Age)_____ (male/female)_____ with a PMHx of _________, presents to this hospital's ER on (date)_____ with (symptoms)_____, (signs)_____ and (investigations)_____. Therefore, this patient has the diagnosis (Dx)_____ and the differential diagnosis (dDx) includes ________________.
15. Problem List: Pneumonic (Every Student Should Attempt to Print Clearly)
    a. (Dx above) - Etiology
       Status
       Severity
       Anatomy
       Physiology
       Complications
       Plan
    b. 2nd problem- Etiology
       Status
       Severity
       Anatomy
       Physiology
       Complications
       Plan
    c. 3rd problem- Etiology
       Status
       Severity
       Anatomy
       Physiology
       Complications
       Plan

End of Dictation

DISCHARGE SUMMARY TEMPLATE

1. Today's Date
2. This is [spell your name] dictating a DC summary for Dr. (Attending's Name)______
3. Patient Name
4. Patient ID number
5. Copies of report to: Fam Dr's name, Attending's name, others
6. Admission Date
7. Discharge Date
8. Discharge Dx:
   1)
   2)
   [new paragraph]
9. Other Diagnoses:
   1)
   etc
   [new paragraph]
10. Discharge Meds: List name route dose freq *indicate meds that have been added, changed or discontinued
    [new paragraph]
11. Course in Hospital (VERY BRIEF!!!)
    (Include only relevant hospital Investigations ie. CXR, ECHO, VQ, CT, US, Endoscopy, Stress Test. EKG, last Hb, Cr, TSH, best FEV1 and best FVC)
    [new paragraph]
12. Follow-up: Appointments, home care, outpatient tests
13. Attending's Name
14. Your Name and status

End of dictation

DEATH

Pronouncing

1. Identify patient by arm tag
2. Non-rousible to verbal/tactile stimuli
3. Ausculate for heart sounds, feel for carotid pulse
4. Look, listen for respiration sounds
5. Record position of pupils reaction to light
6. Record time assessment completed as the time of death
7. Document: Was called to pronounce Mr. X dead. Patient unresponsive to (list #1-6 above).
8. Notify patient's GP and Attending Physician
9. Autopsy??
10. Next of Kin

When to Call the Coroner

The 3 Major Grounds for Reporting a Death

Every death should be considered a potential coroner's case. The decision to contact the coroner is usually made at the time of pronouncement of death, however it could be made following the collection of further information. A coroner should be notified when the death appears to be:

1. Non-Natural:

A death is considered "non-natural" when there is reason to believe that the death may not be entirely due to natural disease. "Natural disease" includes complications of treatment and surgery, provided that death is a known complication of the procedure, and the procedure was for a natural disease.
If it is believed that some external event may have reasonably played a role in the death, a coroner must be notified.
Death usually follows a chain of events or conditions. For instance:
Fall → hip fracture → surgery → post-operative heart attack
By convention, the underlying cause is defined as the primary non-natural event in the sequence (eg. fall) or, if there is none, by the primary natural event.

2. **Specific Circumstances:**

   The law requires that deaths under the following circumstances be reported to the coroner:

- Custody death – while in, or attempting to escape custody, including arrest and detention by police, or while serving a sentence
- Psychiatric hospital inpatients, voluntary or involuntary; whether or not in the hospital at the time of death; and, including a patient transferred from a psychiatric bed to any other hospital inpatient unit
- Maternal death – any mother who dies during pregnancy from any cause whatsoever; or after pregnancy, if the death may be a complication of pregnancy
- Certain Stillbirths– any stillbirth in which the delivery occurs outside hospital or without a physician in attendance.
  (Pursuant to the Vital Statistics Act, a live birth is where the foetus shows any signs of life after delivery, regardless of gestational age or weight. A stillbirth is where the foetus is at least 20 weeks or 500g and shows no signs of life after expulsion. A product of conception is a foetus which is less than 20 weeks gestational age and less than 500g and shows no signs of life after expulsion.)
- Medical Aid in Dying (MAiD) – physician/NP who does MAiD provision should call the coroner after provision complete

3. **Natural deaths with significant issues.**

   These may include, but are not limited to:

- Allegations that the death was preventable, or the result of malpractice or negligence
- Communication and resource issues
- Public safety issues, eg. a death due to infectious disease that may be contagious, such as meningitis.
- "Crib death" – the sudden and unexpected, apparently natural death of a previously well infant under 2 years of age

- Construction and mining deaths
- Some deaths in provincially regulated long-term care facilities such as nursing homes require notification of the coroner. Contact the facility to determine whether or not the case is reportable.
- Potential organ donor who is currently alive but whose death is anticipated and who would be a coroner's case at the time of death

Best Practices for Notification of Coroner

While not required under legislation, it is established best practice to notify the coroner when the following types of death occur:

- Peri-operative: Death during a procedure or before leaving the recovery room, unless death was known to be the most likely outcome prior to the procedure (eg. attempted repair of ruptured AAA in frail elderly person)
- Pediatric deaths
- Workplace deaths
- Deaths of females while alone with intimate male partner

DICTATING

*NOTE - You will have a different dictation number for each of the below sites.

St. Joseph's Healthcare

On-Site Extension: 32078
Off-Site: 905-522-1155 X 32078
Transcription Department Extension: 33822

Telephone Keypad Controls

1. Play
2. Record/Pause
3. Rewind
4. Pause
5. Start second dictation
6. Go to end of dictation
7. Continuous forwarding
8. Go to beginning of dictation
9. Disconnect from system

HHS

On-Site Extension: 5000
Off-site: 905-521-2100 X 5000

Work Type:

1. Inpatient/Emergency Consultation
2. Discharge Summary
3. Operative Report
4. Pre-Anesthesia Clinic
5. Outpatient Clinic
6. Endoscopy

12. Procedure Note

JCC

On-Site Extension: 5000
Off-site: 905-575-6341 X 5000

Work Types:

1. Priority
2. Consult
3. Other

Nausea
1. Gravol
2. ondansetron 4-8mg PO/IV q8hr (QT<450)
3. Domperidone

Sleep
1. Melatonin 3-6mg PO QHS.

Indigestion
1. Tums 1-2tab PO q4hr
2. Ranitidine 150-300mg PO BID.

Gastroparesis
1. Domperidone
2. Metoclopramide.

Arrhythmia
↓HR (ABCD)

Amiodarone 150mg IV bolus over 10min q10-15min or 60mg/hr over 6hr (Max 2.2g/day) → metoprolol tartrate 50-100mg PO BID

β blocker - metoprolol 5mg IV over 1min q5min x3 PRN ↗ bisoprolol 5-10mg PO daily

CCB - diltiazem 15-20mg IV over 2min maintenance 5-20mg/hr IV → diltiazem CD 120-480 PO daily

Digitalis - digoxin 0.25-0.5mg IV q6hr (total dose 1mg) maintenance: 0.125-0.25mg PO/IV daily → digoxin 0.5mg PO 1 dose, 0.25mg x 2 dose, 0.0625-0.25mg daily

Rate control: